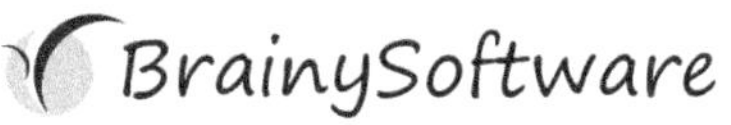

Spring MVC 学习指南

（第2版）

[美] Paul Deck 著
林仪明 译

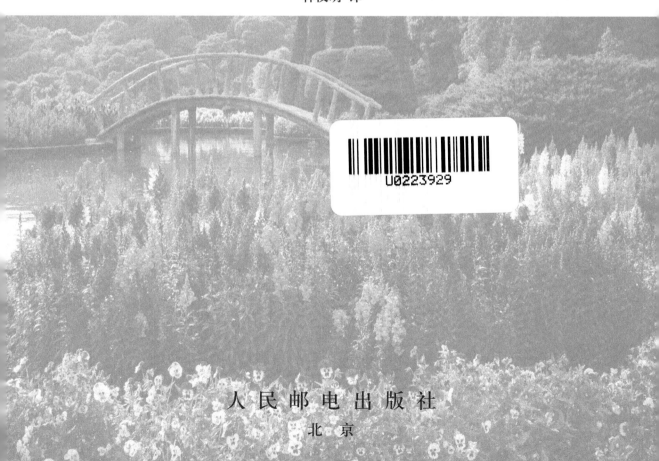

人民邮电出版社
北京

图书在版编目（C I P）数据

Spring MVC学习指南 : 第2版 / (美) 戴克
(Paul Deck) 著 ; 林仪明译. -- 北京 : 人民邮电出版
社, 2017.5(2023.2重印)
ISBN 978-7-115-44759-3

Ⅰ. ①S… Ⅱ. ①戴… ②林… Ⅲ. ①JAVA语言－程序
设计－指南 Ⅳ. ①TP312.8-62

中国版本图书馆CIP数据核字(2017)第031857号

版权声明

◆ 著　　　[美] Paul Deck
译　　　林仪明
责任编辑　陈冀康
责任印制　焦志炜

◆ 人民邮电出版社出版发行　　北京市丰台区成寿寺路 11 号
邮编　100164　　电子邮件　315@ptpress.com.cn
网址　https://www.ptpress.com.cn
固安县铭成印刷有限公司印刷

◆ 开本：800×1000　1/16
印张：21.5　　　2017 年 5 月第 1 版
字数：406 千字　　　2023 年 2 月河北第 12 次印刷
著作权合同登记号　图字：01-2016-5102 号

定价：59.00 元

读者服务热线：(010)81055410　印装质量热线：(010)81055316
反盗版热线：(010)81055315
广告经营许可证：京东市监广登字 20170147 号

内容提要

Spring MVC 是 Spring 框架中用于 Web 应用快速开发的一个模块，其中的 MVC 是 Model-View-Controller 的缩写。作为当今业界最主流的 Web 开发框架，Spring MVC 已经成为当前最热门的开发技能，同时也广泛用于桌面开发领域。

本书重在讲述如何通过 Spring MVC 来开发基于 Java 的 Web 应用。全书共包括 13 章和 5 个附录，分别从 Spring 框架、模型 2 和 MVC 模式、Spring MVC 介绍、控制器、数据绑定和表单标签库、转换器和格式化、验证器、表达式语言、JSTL、国际化、上传文件、下载文件以及应用测试等多个角度介绍了 Spring MVC。附录部分分别介绍了 Tomcat、Spring Tool Suite 和 Maven 等工具的安装和配置，Servlet、JavaServer Pages 和部署描述符的相关参考资料。除此之外，本书还配有丰富的示例以供读者练习和参考。

本书是一本 Spring MVC 的教程，内容细致、讲解清晰，非常适合 Web 开发者和想要使用 Spring MVC 开发基于 Java 的 Web 应用的读者阅读。

作者简介

Paul Deck 是一位资深的 Java 和 Spring 专家，开发过大型的企业级应用，并且目前是一位独立的软件工程师。他是 《How Tomcat Works》一书的合著者。

译者简介

林仪明，男，现为 IBM 高级工程师。2004 年毕业于厦门大学软件学院，主要研究软件架构、应用中间件。目前在福州生活和工作，先后从事软件技术开发，软件架构设计以及团队管理等工作，有多年的开发设计和管理经验，目前提供 IBM 中间件产品支持工作。

前言

欢迎阅读本书。

Spring MVC 是 Spring 框架中用于 Web 应用快速开发的一个模块。Spring MVC 的 MVC 是 Model-View-Controller 的缩写。它是一个广泛应用于图形化用户交互开发中的设计模式，不仅常见于 Web 开发，也广泛应用于如 Swing 和 JavaFX 等桌面开发。

作为当今业界最主流的 Web 开发框架，Spring MVC（有时也称 Spring Web MVC）的开发技能相当热门。本书可供想要学习如何通过 Spring MVC 开发基于 Java 的 Web 应用的开发人员阅读。

Spring MVC 基于 Spring 框架、Servlet 和 JSP（JavaServer Page），在掌握这 3 种技术的基础上学习 Spring MVC 将非常容易。本书第 1 章针对 Spring 新手提供一个快速教程，附录 C 和附录 D 将帮助你快速学习 Servlet 和 JSP。如果希望深入学习 Servlet 和 JSP，推荐阅读由 Budi Kurniawan 所著的《*Servlet and JSP：A Tutorial，second Edition*》。

接下来，我们将介绍 HTTP、基于 Servlet 和 JSP 的 Web 编程，以及本书的章节结构。

HTTP

HTTP 使得 Web 服务器与浏览器之间可以通过互联网或内网进行数据交互。作为一个制定标准的国际社区，万维网联盟（W3C）负责和维护 HTTP。HTTP 第一版是 HTTP 0.9，随后更新为 HTTP 1.0，之后的版本是 HTTP 1.1。HTTP 1.1 版本的 RFC 编号是 2616。编写本书时，HTTP 1.1 依然是当前最流行的 HTTP 版本。当前最新版本是发布于 2015 年 5 月的 HTTP/2，附表 1.1 列出了 HTTP 各个版本及其发布时间，HTTP 的第二个主要版本通常表示为 HTTP/2，而不是 HTTP2。

附表 1.1　HTTP 版本与发布日期

版本	发布时间
HTTP 0.9	1991 年
HTTP 1.0	1996 年
HTTP 1.1	1997 年发布，1999 年更新
HTTP/2	2015 年 5 月

Web 服务器每天 24 小时不间断运行，并等待 HTTP 客户端（通常是 Web 浏览器）来连接并请求资源。通常，客户端发起一个连接，服务端不会主动连接客户端。

互联网用户需要通过点击链接或者输入一个 URL 地址来访问一个资源，以下为两个示例：

```
http://google.com/index.html
http://facebook.com/index.html
```

URL 的第一个部分是 HTTP，代表所采用的协议。除 HTTP 外，URL 还可以采用其他类型的协议，以下为两个示例：

```
mailto:joe@example.org
ftp://marketing@ftp.example.org
```

通常，HTTP 的 URL 格式如下：

protocol://[host.]domain[:port][/context][/resource][?query string | path variable]

或者

protocol://IP Address[:port][/context][/resource][?query string | path variable]

中括号中的内容是可选项。因此，一个最简单的 URL 是 http://yahoo.ca 或者是 http://192.168.1.9。

需要说明的是，除了输入 http://google.com 外，还可以用 http://173.194.46.35 来访问谷歌。可以用 ping 命令来获取域名对应的 IP 地址。

```
ping google.com
```

由于 IP 地址不容易记忆，所以实践中更倾向于使用域名。一台计算机可以托管不止一个域名，因此，不同的域名可能指向同一个 IP。另外，example.com 或者 example.org 无法被注册，因为它们被保留作为各类文档手册举例使用。

URL 中的 host 部分用来表示在互联网或内网中一个唯一的地址。例如，http://yahoo.com（没有 host）访问的地址完全不同于 http://mail.yahoo.com（有 host）。多年以来，作为最受欢迎的主机名，www 是默认的主机名。通常，http://www.domainName 会被映射到 http://domainName。

HTTP 的默认端口是 80 端口。因此，对于采用 80 端口的 Web 服务器，无需输入端口号。有时，Web 服务器并未运行在 80 端口上，此时必须输入相应的端口号。例如，Tomcat 服务器的默认端口号是 8080，为了能正确访问服务器，必须提供输入端口号。

```
http://localhost:8080/index.html
```

localhost 作为一个保留关键字，用于指向本机。

URL 中的 context 部分用来代表应用名称，该部分也是可选的。一台 Web 服务器可以运行多个上下文（应用），其中一个可以配置为默认上下文。若访问默认上下文中的资源，可以跳过 context 部分。

最后，一个 context 可以有一个或多个默认资源（通常为 index.html、index.htm 或者 default.htm）。一个不带资源名称的 URL 通常指向默认资源。当存在多个默认资源时，其中最高优先级的资源将被返回给客户端。

资源名后可以有一个或多个查询语句或者路径参数。查询语句是一个 key/value 组，多个查询语句间用“&”符号分隔。路径参数类似于查询语句，但只有 value 部分，多个 value 部分用“/”符号分隔。

接下来，我们介绍 HTTP 请求和响应。

HTTP 请求

一个 HTTP 请求包含 3 部分内容。

1．方法—URI—协议/版本。

2．请求头信息。

3．请求正文。

下面为一个 HTTP 请求示例。

```
POST /examples/default.jsp HTTP/1.1
Accept: text/plain; text/html
Accept-Language:en-gb
Connection:Keep-Alive
Host:localhost
User-Agent: Mozilla/5.0 (Macintosh; U; Intel Mac OS X 10.5; en-US; rv:1.9.2.6) Gecko/20100625 Firefox/3.6.6
Content-Length:30
Content-Type: application/x-www-form-urlencoded
Accept-Encoding: gzip, deflate

lastName=Blanks&firstName=Mike
```

请求的第一行 POST /examples/default.jsp HTTP/1.1 是方法—URI—协议/版本。

请求方法为 POST，URI 为/examples/default.jsp，而协议/版本为 HTTP/1.1。

HTTP 1.1 规范定义了 7 种类型的方法，包括 GET、POST、HEAD、OPTIONS、PUT、DELETE 以及 TRACE，其中 GET 和 POST 广泛应用于互联网。

URI 定义了一个互联网资源，通常解析为服务器根目录的相对路径。因此，通常用“/”符号打头。另外，URL 是 URI 的一个具体类型（详见 http://www.ietf.org/rfc/rfc2396.txt）。

HTTP 请求包含的请求头信息，包括关于客户端环境以及实体内容等非常有用的信息。例如，浏览器设置的语言、实体内容长度等。每个 header 都用回车/换行（即 CRLF）分隔。

HTTP 请求头信息和请求正文用一行空行分隔，HTTP 服务器据此判断请求正文的起始位置。因此，在一些关于互联网的书籍中，CRLF 被作为 HTTP 请求的第 4 种组件。

示例中，请求正文是 lastName=Blanks&firstName=Mike。

在正常的 HTTP 请求中，请求正文的内容不止如此。

HTTP 响应

同 HTTP 请求一样，HTTP 响应也包含 3 部分。

1．协议—状态码—描述。

2．响应头信息。

3．响应正文。

下面为一个 HTTP 响应示例。

```
HTTP/1.1 200 OK
Server: Apache-Coyote/1.1
Date: Thu, 29 Sep 2013 13:13:33 GMT
Content-Type: text/html
Last-Modified: Web, 28 Sep 2013 13:13:12 GMT
Content-Length: 112

<html>
<head>
<title>HTTP Response Example</title>
</head>
<body>
Welcome to Brainy Software
</body>
</html>
```

类似于 HTTP 请求报文，HTTP 响应报文的第一行说明了 HTTP 的版本是 1.1，并且请求结果是成功的（状态代码 200 为响应成功）。

同 HTTP 请求报文头信息一样，HTTP 响应报文头信息也包含了大量有用的信息。HTTP 响应报文的响应正文是 HTML 文档。HTTP 响应报文的头信息和响应正文也是用 CRLF 分隔的。

状态代码 200 表示 Web 服务器能正确响应所请求的资源。若一个请求的资源不能被找到或者理解，则 Web 服务器将返回不同的状态代码。例如，访问未授权的资源将返回 401，而使用被禁用的请求方法将返回 405。完整的 HTTP 响应状态代码列表详见网址 http://www.w3.org/Protocols/rfc2616/rfc2616-sec10.html。

Servlet 和 JSP

Java Servlet 技术是 Java 体系中用于开发 Web 应用的底层技术。1996 年，Servlet 和 JSP 由 SUN 系统公司发布，以替代 CGI 技术，作为产生 Web 动态内容的标准。CGI 技术为每一个请求创建相应的进程，但是，创建进程会耗费大量的 CPU 周期，最终导致很难编写可伸缩的 CGI 程序。相对于 CGI 程序，一个 Servlet 则快多了，这是因为当一个 Servlet 为响应第一次请求而创建后，会驻留在内存中，以便响应后续请求。

从 Servlet 技术出现的那天起，人们开发了大量的 Web 框架来帮助程序员快速编写 Web 应用程序。这些开发框架让开发人员能更关注业务逻辑，减少编写“相似”的代码片段。尽管如此，开发人员依然需要去理解 Servlet 技术的基础概念。随后发布的 JavaServer Pages（JSP）技术，是用来帮助简化 Servlet 开发。尽管实践中会应用一些诸如 Spring MVC、Struct 2 或者 JSF 等强大的开发框架，但如果没有深入理解 Servlet 和 JSP 技术，则无法有效和高效地开发。Servlet 是运行在 Servlet 容器中的 Java 程序，而 Servlet 容器或 Servlet 引擎相当于一个 Web 服务器，但是可以产生动态内容，而不仅是静态资源。Servlet 当前的版本为 3.1，其规范定义可见 JSR（Java Specification Request）340（http://jcp.org/en/jsr/detail?id=340），基于 Java 标准版本 6 及以上版本。JSP 2.3 规范定义在 JSR 245（http://jcp.org/en/jsr/detail?id=245）。本书假定读者已经了解 Java 以及面向对象编程技术。对于 Java 新手，推荐阅读《Java 7：A Beginner’s Tutorial（fourth Edition)》。

一个 Servlet 是一个 Java 程序，一个 Servlet 应用包含了一个或多个 Servlet，一个 JSP 页面会被翻译并编译成一个 Servlet。

一个 Servlet 应用运行在一个 Servlet 容器中，它无法独立运行。Servlet 容器将来自用户的请求传递给 Servlet 应用，并将 Servlet 应用的响应返回给用户。由于大部分 Servlet 应用都会包含一些 JSP 界面，故称 Java Web 应用为“Servlet/JSP”应用会更恰当些。

Web 用户通过一个诸如 IE、火狐或者 Chrome 等 Web 浏览器来访问 Servlet 应用。Web 浏览器又称为 Web 客户端。图 1 展示了一个典型的 Servlet/JSP 应用架构。

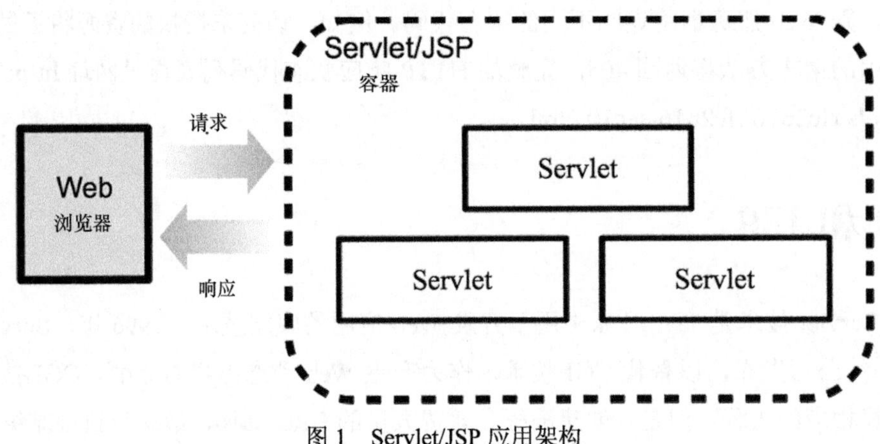

图 1　Servlet/JSP 应用架构

Web 服务端和 Web 客户端基于 HTTP 通信。因此，Web 服务端往往也称为 HTTP 服务端。

一个 Servlet/JSP 容器是一个能处理 Servlet 以及静态资源的 Web 服务端。过去，由于 HTTP 服务器更加健壮，人们更愿意将 Servlet/JSP 容器作为 HTTP 服务器（如 Apache HTTP 服务器）的一个模块来运行，在这种场景下，Servlet/JSP 容器用来产生动态内容，而 HTTP 服务器处理静态资源。今天，Servlet/JSP 容器已经更加成熟，并且被广泛地独立部署。Apache Tomcat 和 Jetty 作为最流行的 Servlet/JSP 容器，免费而且开源。下载地址为 http://tomcat.apache.org 和 http://jetty.codehaus.org。

Servlet 和 JSP 仅是 Java 企业版众多技术之一，其他 Java 企业版技术包括 JMS、EJB、JSF 和 JPA 等。Java 企业版 7（当前最新版）完整的技术列表见 Http://www.oracle.com/technetwork/java/javaee/tech/index.html。

运行一个 Java 企业版应用需要一个 Java 企业版容器，常见的容器有 GlassFish、JBoss、Oracle Weblogic 以及 IBM WebSphere。虽然可以将一个 Servlet/JSP 应用部署到 Java 企业版容器中，但一个 Servlet/JSP 容器其实就足够了，而且更加轻量级。Tomcat 和 Jetty 不是 Java 企业版容器，故它们不能运行 EJB 或 JMS。

下载 Spring 或使用 STS 与 Maven/Gradle

Spring 网站建议使用 Maven 和 Gradle 来下载 Spring 库及其依赖包。现在的应用通常有很多依赖包，并且这些依赖包也有自己的依赖。通过一个依赖管理工具，可以把我们从解析和下载依赖的工作中解放出来。并且这些工具还可以帮助构建、测试和打包应用。Maven 和 Gradle 是当下流行的依赖管理系统，使用这两个工具是一个明智的选择。然而当下还有很多 Java 开发者要么不熟悉这两个工具，要么不喜欢使用或者不想使用这两个工具。一个好的教程不能要求他们先学习 Maven 或 Gradle，然后才能构建和测试他们的 Spring 应用程序。因此，本书所有的例子都有两种发行包：基于 Maven 的发行包和不依赖 Maven/Gradle 的发行包。

Maven 曾经是 Spring 支持的唯一的依赖管理系统。Spring 开发工具 Spring Tool Suites（STS），是一个基于 Eclipse 的 IDE，捆绑了一个最新的 Maven 版本。这对于不熟悉 Maven 的开发人员是一个好消息，只要熟悉 Eclipse，则无须学习 Maven 就可以应用其好处。然而，在写作本书的时候，最新版本的 STS 并不是没有缺点。例如，当使用 STS 创建 Spring MVC 应用程序时，生成的应用程序会因为缺少 Servlet 部署描述符（web.xml 文件）而导致 Maven 插件生成一个错误消息。因此开发人员不得不被迫处理一个 Maven 问题。当然，可以通过简单地创建一个 web.xml 文件来解决它。但是，从 Servlet 3.0 开始，web.xml 文件是可选的，并且可以构建一个没有 web.xml 文件的 Spring MVC 应用程序。

下面对于使用 Maven/Gradle 或不使用工作会分别介绍。

手动下载 Spring

如果不想通过 Gradle 和 Maven 来构建应用，则应从 Spring 仓库中下载其类库：

```
http://repo.spring.io/release/org/springframework/spring/
```

点击以上链接将跳转到最新版本（写作本书时为版本为 4.2.5），选择一个 zip 发行包，发行包命名格式如下：

```
spring-framework-x.y.z.RELEASE-dist.zip
```

其中，x.y.z 表示 Spring 的主要版本和次要版本。

将下载好的 zip 解压到任意目录。在解压的目录中，包含相应的文档和 Java 源代码，其中 libs 文件夹下为基于 Spring 框架开发应用所需的 jar 文件。

使用 STS 和 Maven/Gradle

如果选择使用 Maven 或 Gradle，则无需手动下载 Spring。具体内容，请参考本书附录 B。

下载 Spring 源码

Spring 框架是一个开源项目，如果你喜欢冒险或者你想要的尚未发布的最新版本的 Spring，你可以使用 Git 下载源代码。克隆 Spring 源代码的命令如下：

```
git clone git://github.com/spring-projects/spring-framework.git
```

此外还需使用 Gradle 来从源代码构建 Spring。Git 和 Gradle 工具不是本书的范围。

本书内容简介

第 1 章：Spring 框架，介绍了最流行的开源框架。

第 2 章：模型 2 和 MVC 模式，讨论了 Spring MVC 实现的设计模式。

第 3 章：Spring MVC 介绍，编写了第一个 Spring MVC 应用。

第 4 章：基于注解的控制器，讨论了 MVC 模式中最重要的一个对象——控制器。在本章中，我们将学习如何编写基于注解的控制器，该方式由 Spring MVC 2.5 版本引入。

第 5 章：数据绑定和表单标签库，讨论 Spring MVC 最强大的一个特性，并利用它来展示表单数据。

第 6 章：转换器和格式化，讨论了数据绑定的辅助对象类型。

第 7 章：验证器，展示如何通过验证器来验证用户输入数据。

第 8 章：表达式语言，介绍了 JSP 2.0 中最重要的特性"表达式语言"。该特性的目标是帮助开发人员编写无脚本的 JSP 页面，让 JSP 页面更加简洁而且有效。本章将帮助你学会通过 EL 来访问 Java Bean 和上下文对象。

第 9 章：JSTL，介绍了 JSP 技术中最重要的类库——标准标签库。这是一组帮助处理常见问题的标签，诸如访问 map 或集合对象、条件判断、XML 处理，以及数据库访问和数据处理。

第 10 章：国际化，将展示如何用 Spring MVC 来构建多语言网站。

第 11 章：上传文件，介绍两种不同的方式来处理文件上传。

第 12 章：下载文件，介绍如何用编程方式向客户端端传输一个资源。

第 13 章：应用测试，介绍如何使用 Junit、Mockito 和 Spring MVC Test 进行单元测试和集成测试。

附录 A：Tomcat，介绍如何安装和配置 Tomcat。

附录 B： Spring Tool Suite 和 Maven，介绍如何安装并配置 Spring Tool Suite，并且开发和运行 Spring MVC 应用。

附录C：Servlet，介绍了Servlet API并展示几个简单的Servlet应用。本附录重点关注Servlet API的4个包中的两个，即javax.servlet和javax.servlet.http。

附录D：JavaServer Pages，介绍了JSP语法，包括指令、脚本元素和actions。

附录E：部署描述符，介绍如何配置Spring MVC应用以便部署。

下载示例应用

本书所有的示例应用压缩包可以通过以下地址下载：
http://books.brainysoftware.com/download 和 http://www.epubit.com.cn。

目　录

第 1 章

Spring 框架

Spring 框架是一个开源的企业应用开发框架，作为一个轻量级的解决方案，它包含 20 多个不同的模块。本书主要关注 Core、Spring Bean、Spring MVC 和 Spring MVC Test 模块。

本章主要介绍 Core 和 Spring Bean 这两个模块，以及它们如何提供依赖注入解决方案。为方便初学者，本书会深入讨论依赖注入概念的细节。后续介绍开发 MVC 应用的章节将会使用到本章介绍的技能。

依赖注入

在过去数年间，依赖注入技术作为代码可测试性的一个解决方案已经被广泛应用。实际上，Spring、谷歌 Guice 等框架都采用了依赖注入技术。那么，什么是依赖注入技术？

很多人在使用中并不区分依赖注入和控制反转（IoC），尽管 Martin Fowler 在其文章中已分析了两者的不同：

```
http://martinfowler.com/articles/injection.html
```

简单来说，依赖注入的情况如下。

有两个组件 A 和 B，A 依赖于 B。假定 A 是一个类，且 A 有一个方法 importantMethod 用到了 B，如下：

```
public class A {
    public void importantMethod() {
        B b = ... // get an instance of B
        b.usefulMethod();
        ...
    }
    ...
}
```

要使用 B，类 A 必须先获得组件 B 的实例引用。若 B 是一个具体类，则可通过 new 关键字直接创建组件 B 实例。但是，如果 B 是接口，且有多个实现，则问题就变得复杂了。我们固然可以任意选择接口 B 的一个实现类，但这也意味着 A 的可重用性大大降低了，因为无法采用 B 的其他实现。

依赖注入是这样处理此类情景的：接管对象的创建工作，并将该对象的引用注入需要该对象的组件。以上述情况为例，依赖注入框架会分别创建对象 A 和对象 B，将对象 B 注入到对象 A 中。

为了能让框架进行依赖注入，程序员需要编写特定的 set 方法或者构建方法。例如，为了能将 B 注入到 A 中，类 A 会被修改成如下形式：

```
public class A {
    private B b;
    public void importantMethod() {
        // no need to worry about creating B anymore
        // B b = ... // get an instance of B
        b.usefulMethod();
        ...
    }
    public void setB(B b) {
        this.b = b;
    }
}
```

修改后的类 A 新增了一个 set 方法，该方法将会被框架调用，以注入 B 的一个实例。由于对象依赖由依赖注入，类 A 的 importantMethod 方法不再需要在调用 B 的 usefulMethod 方法前去创建 B 的一个实例。

当然，也可以采用构造器方式注入，如下所示：

```
public class A {
    private B b;

    public A(B b) {
        this.b = b;
    }

    public void importantMethod() {
        // no need to worry about creating B anymore
        // B b = ... // get an instance of B
        b.usefulMethod();
        ...
    }
}
```

本例中，Spring 会先创建 B 的实例，再创建实例 A，然后把 B 注入到实例 A 中。

注：

Spring 管理的对象称为 beans。

通过提供一个控制反转容器（或者依赖注入容器），Spring 为我们提供一种可以“聪明”地管理 Java 对象依赖关系的方法。其优雅之处在于，程序员无需了解 Spring 框架的存在，更不需要引入任何 Spring 类型。

从 1.0 版本开始，Spring 就同时支持 setter 和构造器方式的依赖注入。从 2.5 版本开始，通过 Autowired 注解，Spring 支持基于 field 方式的依赖注入，但缺点是程序必须引入 org.springframework.beans.factory.annotation.Autowired，这对 Spring 产生了依赖，这样，程序无法直接迁移到另一个依赖注入容器间。

使用 Spring，程序几乎将所有重要对象的创建工作移交给 Spring，并配置如何注入依赖。Spring 支持 XML 或注解两种配置方式。此外，还需要创建一个 ApplicationContext 对象，代表一个 Spring 控制反转容器，org.springframework.context.ApplicationContext 接口有多个实现，包括 ClassPathXmlApplicationContext 和 FileSystemXmlApplicationContext。这两个实现都需要至少一个包含 beans 信息的 XML 文件。ClassPathXmlApplicationContext 尝试在类加载路径中加载配置文件，而 FileSystemXmlApplicationContext 则从文件系统中加载。

下面是从类路径中加载 config1.xml 和 config2.xml 的 ApplicationContext 创建的一个代码示例。

```
ApplicationContext context = new ClassPathXmlApplicationContext(
        new String[] {"config1.xml", "config2.xml"});
```

可以通过调用 ApplicationContext 的 getBean 方法获得对象。

```
Product product = context.getBean("product", Product.class);
```

getBean 方法会查询 id 为 product 且类型为 Product 的 bean 对象。

注：

理想情况下，我们只需在测试代码中创建一个 ApplicationContext，应用程序本身无需处理。对于 Spring MVC 应用，可以通过一个 Spring Servlet 来处理 ApplicationContext，而无需直接处理。

1.1　XML 配置文件

从 1.0 版本开始，Spring 就支持基于 XML 的配置；从 2.5 版本开始，增加了通过注解的配置支持。下面介绍如何配置 XML 文件。配置文件的根元素通常为 beans：

```
<?xml version="1.0" encoding="UTF-8"?>
<beans xmlns="http://www.springframework.org/schema/beans"
  xmlns:xsi="http://www.w3.org/2001/XMLSchema-instance"
  xsi:schemaLocation="http://www.springframework.org/schema/beans
  http://www.springframework.org/schema/beans/spring-beans.xsd">

    ...
</beans>
```

如果需要更强的 Spring 配置能力，可以在 schema location 属性中添加相应的 schema，也可以指定 schema 版本：

```
http://www.springframework.org/schema/beans/spring-beans-4.2.xsd
```

不过，笔者推荐使用默认 schema，以便升级 spring 库时无需修改配置文件。

配置文件既可以是一份，也可以分解为多份，以支持模块化配置。ApplicationContext 的实现类支持读取多份配置文件。另一种选择是，通过一份主配置文件，将该文件导入到其他配置文件。

下面是导入其他配置文件的一个示例：

```
<?xml version="1.0" encoding="UTF-8"?>
<beans xmlns="http://www.springframework.org/schema/beans"
  xmlns:xsi="http://www.w3.org/2001/XMLSchema-instance"
  xsi:schemaLocation="http://www.springframework.org/schema/beans
  http://www.springframework.org/schema/beans/spring-beans.xsd">

    <import resource="config1.xml"/>
    <import resource="module2/config2.xml"/>
    <import resource="/resources/config3.xml"/>
  ...
</beans>
```

bean 元素的配置后面将会详细介绍。

1.2　Spring 控制反转容器的使用

本节主要介绍 Spring 如何管理 bean 和依赖关系。

1.2.1　通过构造器创建一个 bean 实例

前面已经介绍，通过调用 ApplicationContext 的 getBean 方法可以获取一个 bean 的实例。

下面的配置文件中定义了一个名为 product 的 bean（见清单 1.1）。

清单 1.1 一个简单的配置文件

```
<?xml version="1.0" encoding="UTF-8"?>
<beans xmlns="http://www.springframework.org/schema/beans"
  xmlns:xsi="http://www.w3.org/2001/XMLSchema-instance"
  xsi:schemaLocation="http://www.springframework.org/schema/beans
  http://www.springframework.org/schema/beans/spring-beans.xsd">

    <bean name="product" class="springintro.bean.Product"/>

</beans>
```

该 bean 的定义告诉 Spring，通过默认无参的构造器来初始化 Product 类。如果不存在该构造器（如果类的编写者重载了构造器，且没有显示声明默认构造器），则 Spring 将抛出一个异常。此外，该无参数的构造器并不要求是 public 签名。

注意，应采用 id 或者 name 属性标识一个 bean。为了让 Spring 创建一个 Product 实例，应将 bean 定义的 name 值“product”（具体实践中也可以是 id 值）和 Product 类型作为参数传递给 ApplicationContext 的 getBean 方法。

```
ApplicationContext context =
        new ClassPathXmlApplicationContext(
        new String[] {"spring-config.xml"});
Product product1 = context.getBean("product", Product.class);
product1.setName("Excellent snake oil");
System.out.println("product1: " + product1.getName());
```

1.2.2 通过工厂方法创建一个 bean 实例

大部分类可以通过构造器来实例化。然而，Spring 还同样支持通过调用一个工厂的方法来初始化类。

下面的 bean 定义展示了通过工厂方法来实例化 java.time.LocalDate。

```
<bean id="localDate" class="java.time.LocalDate"
    factory-method="now"/>
```

本例中采用了 id 属性而非 name 属性来标识 bean，采用了 getBean 方法来获取 LocalDate 实例。

```
ApplicationContext context =
        new ClassPathXmlApplicationContext(
```

```
        new String[] {"spring-config.xml"});
LocalDate localDate = context.getBean("localDate", LocalDate.class);
```

1.2.3　销毁方法的使用

有时，我们希望一些类在被销毁前能执行一些方法。Spring 考虑到了这样的需求。可以在 bean 定义中配置 destroy-method 属性，来指定在销毁前要执行的方法。

下面的例子中，我们配置 Spring 通过 java.util.concurrent.Executors 的静态方法 newCachedThreadPool 来创建一个 java.uitl.concurrent.ExecutorService 实例，并指定了 destroy-method 属性值为 shutdown 方法。这样，Spring 会在销毁 ExecutorService 实例前调用其 shutdown 方法。

```
<bean id="executorService" class="java.util.concurrent.Executors"
    factory-method="newCachedThreadPool"
    destroy-method="shutdown"/>
```

1.2.4　向构造器传递参数

Spring 支持通过带参数的构造器来初始化类（见清单 1.2）。

清单 1.2　Product 类

```
package springintro.bean;
import java.io.Serializable;

public class Product implements Serializable {
    private static final long serialVersionUID = 748392348L;
    private String name;
    private String description;
    private float price;

    public Product() {
    }

    public Product(String name, String description, float price) {
        this.name = name;
        this.description = description;
        this.price = price;
    }
    public String getName() {
        return name;
    }
    public void setName(String name) {
        this.name = name;
```

```
    }
    public String getDescription() {
        return description;
    }
    public void setDescription(String description) {
        this.description = description;
    }
    public float getPrice() {
        return price;
    }
    public void setPrice(float price) {
        this.price = price;
    }
}
```

以下的定义展示了如何通过参数名传递参数。

```
<bean name="featuredProduct" class="springintro.bean.Product">
    <constructor-arg name="name" value="Ultimate Olive Oil"/>
    <constructor-arg name="description"
        value="The purest olive oil on the market"/>
    <constructor-arg name="price" value="9.95"/>
</bean>
```

这样，在创建 Product 实例时，Spring 会调用如下构造器：

```
public Product(String name, String description, float price) {
    this.name = name;
    this.description = description;
    this.price = price;
}
```

除了通过名称传递参数外，Spring 还支持通过指数方式来传递参数，具体如下：

```
<bean name="featuredProduct2" class="springintro.bean.Product">
    <constructor-arg index="0" value="Ultimate Olive Oil"/>
    <constructor-arg index="1"
        value="The purest olive oil on the market"/>
    <constructor-arg index="2" value="9.95"/>
</bean>
```

需要说明的是，采用这种方式，对应构造器的所有参数必须传递，缺一不可。

1.2.5 Setter 方式依赖注入

下面以 Employee 类和 Address 类为例，介绍 setter 方式依赖注入（见清单 1.3 和清单 1.4）。

清单 1.3 Employee 类

```
package springintro.bean;

public class Employee {
    private String firstName;
    private String lastName;
    private Address homeAddress;

    public Employee() {
    }

    public Employee(String firstName, String lastName, Address
       homeAddress) {
        this.firstName = firstName;
        this.lastName = lastName;
        this.homeAddress = homeAddress;
    }

    public String getFirstName() {
        return firstName;
    }

    public void setFirstName(String firstName) {
        this.firstName = firstName;
    }

    public String getLastName() {
        return lastName;
    }

    public void setLastName(String lastName) {
        this.lastName = lastName;
    }

    public Address getHomeAddress() {
        return homeAddress;
    }

    public void setHomeAddress(Address homeAddress) {
        this.homeAddress = homeAddress;
    }

    @Override
    public String toString() {
        return firstName + " " + lastName
```

```
                + "\n" + homeAddress;
    }
}
```

清单 1.4 Address 类

```
package springintro.bean;

public class Address {
  private String line1;
    private String line2;
    private String city;
    private String state;
    private String zipCode;
    private String country;

    public Address(String line1, String line2, String city,
            String state, String zipCode, String country) {
        this.line1 = line1;
        this.line2 = line2;
        this.city = city;
        this.state = state;
        this.zipCode = zipCode;
        this.country = country;
    }

    // getters and setters omitted

    @Override
    public String toString() {
        return line1 + "\n"
                + line2 + "\n"
                + city + "\n"
                + state + " " + zipCode + "\n"
                + country;
    }
}
```

Employee 依赖于 Address 类，可以通过如下配置来保证每个 Employee 实例都能包含 Address 实例。

```
<bean name="simpleAddress" class="springintro.bean.Address">
    <constructor-arg name="line1" value="151 Corner Street"/>
    <constructor-arg name="line2" value=""/>
    <constructor-arg name="city" value="Albany"/>
    <constructor-arg name="state" value="NY"/>
    <constructor-arg name="zipCode" value="99999"/>
```

```
    <constructor-arg name="country" value="US"/>
</bean>

<bean name="employee1" class="springintro.bean.Employee">
    <property name="homeAddress" ref="simpleAddress"/>
    <property name="firstName" value="Junior"/>
    <property name="lastName" value="Moore"/>
</bean>
```

simpleAddress 对象是 Address 类的一个实例，它通过构造器方式实例化。employee1 对象则通过配置 property 元素来调用 setter 方法以设置值。需要注意的是，homeAddress 属性配置的是 simpleAddress 对象的引用。

被引用对象的配置定义无需早于引用其对象的定义。在本例中，employee1 对象可以出现在 simpleAddress 对象定义之前。

1.2.6　构造器方式依赖注入

清单 1.3 所示的 Employee 类提供了一个可以传递参数的构造器，我们还可以将 Address 对象通过构造器注入，如下所示：

```
<bean name="employee2" class="springintro.bean.Employee">
    <constructor-arg name="firstName" value="Senior"/>
    <constructor-arg name="lastName" value="Moore"/>
    <constructor-arg name="homeAddress" ref="simpleAddress"/>
</bean>

<bean name="simpleAddress" class="springintro.bean.Address">
    <constructor-arg name="line1" value="151 Corner Street"/>
    <constructor-arg name="line2" value=""/>
    <constructor-arg name="city" value="Albany"/>
    <constructor-arg name="state" value="NY"/>
    <constructor-arg name="zipCode" value="99999"/>
    <constructor-arg name="country" value="US"/>
</bean>
```

1.3　小结

本章学习了依赖注入的概念以及基于 Spring 容器的实践，后续将在此基础之上配置 Spring 应用。

第 2 章 模型 2 和 MVC 模式

Java Web 应用开发中有两种设计模型，为了方便，分别称为模型 1 和模型 2。模型 1 是以页面中心，适合于小应用开发。而模型 2 基于 MVC 模式，是 Java Web 应用的推荐架构（简单类型的应用除外）。

本章将会讨论模型 2，并展示 4 个不同示例应用。第一个应用是一个基本的模型 2 应用，采用 Servlet 作为控制器；第二个应用也是模型 2 应用，但采用了 Filter 作为控制器；第三个应用引入了验证控件来校验用户的输入；最后一个应用则采用了一个自研的依赖注入器。在实践中，应替换为 Spring。

注

在写作本书时，业界正致力于 MVC Web 框架的标准化（见 JSR 371）。

2.1 模型 1 介绍

第一次学习 JSP，通常通过链接方式进行 JSP 页面间的跳转。这种方式非常直接，但在中型和大型应用中，这种方式会带来维护上的问题。修改一个 JSP 页面的名字，会导致页面中大量的链接需要修正。因此，实践中并不推荐模型 1（但仅 2~3 个页面的应用除外）。

2.2 模型 2 介绍

模型 2 基于模型—视图—控制器（MVC）模式，该模式是 Smalltalk-80 用户交互的核心概念，那时还没有设计模式的说法，当时称为 MVC 范式。

一个实现 MVC 模式的应用包含模型、视图和控制器 3 个模块。视图负责应用的展示。模型封装了应用的数据和业务逻辑。控制器负责接收用户输入，改变模型以及调整视图的显示。

模型 2 中，Servlet 或者 Filter 都可以充当控制器。几乎所有现代 Web 框架都是模型 2 的实现。Struts 1、Spring MVC 和 JavaServer Faces 使用一个 Servlet 作为控制器，而 Struts 2 则使用一个 Filter 作为控制器。大部分都采用 JSP 页面作为应用的视图，当然也有其他技术。而模型则采用 POJO（Plain Old Java Object）。不同于 EJB 等特定对象，POJO 是一个普通对象。实践中会采用一个 JavaBean 来持有模型状态，并将业务逻辑放到一个 Action 类中。

图 2.1 展示了一个模型 2 应用的架构图。

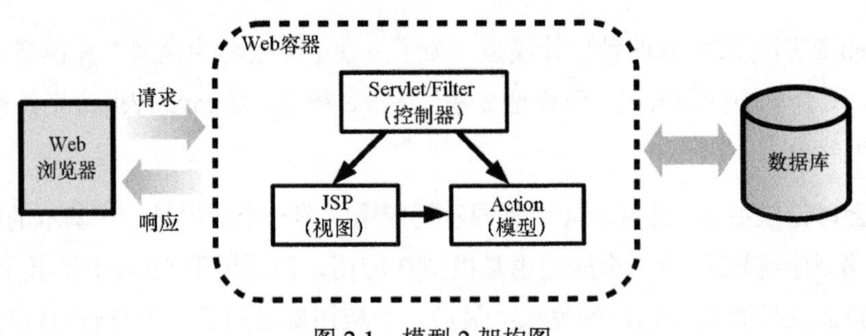

图 2.1　模型 2 架构图

每个 HTTP 请求都发送给控制器，请求中的 URI 标识出对应的 action。action 代表了应用可以执行的一个操作。一个提供了 action 的 Java 对象称为 action 对象。一个 action 类可以支持多个 action（在 Spring MVC 以及 Struts 2 中），或者一个 action（在 Struts 1 中）。

看似简单的操作可能需要多个 action。如向数据库添加一个产品，需要两个 action。

（1）显示一个“添加产品”的表单，以便用户能输入产品信息。

（2）将表单信息保存到数据库中。

如前所述，我们需要通过 URI 方式告诉控制器执行相应的 action。例如，通过发送类似如下的 URI，来显示“添加产品”表单。

```
http://domain/appName/input-product
```

通过类似如下的 URI，来保存产品。

```
http://domain/appName/save-product
```

控制器会解析 URI 并调用相应的 action，然后将模型对象放到视图可以访问的区域（以便服务端数据可以展示在浏览器上）。最后，控制器利用 RequestDispatcher 或者 HttpServlet Response.sendRedirect()方法跳转到视图（JSP 页面或者其他资源）。在 JSP 页面中，用表达式语言以及定制标签显示数据。

注意

调用 RequestDispatcher.forward 方法或者 HttpServletResponse.sendRedirect()方法并不会停止执行剩余的代码。因此，若 forward 方法不是最后一行代码，则应显式地返回。

```
if (action.equals(...)) {
    RequestDispatcher rd = request.getRequestDispatcher(dispatchUrl);
    rd.forward(request, response);
    return;//explicitly return. Or else, the code below will be executed
}
// do something else
```

大多数时候，你将使用 RequestDispatcher 转发到视图，因为它比 sendRedirect 更快响应。这是因为重定向导致服务器向浏览器发送状态代码为 302 的 HTTP 响应，并包含新 URL。 而浏览器在接收到状态代码 302 时，根据响应头部中找到的 URL 向服务器发出新的 HTTP 请求。换句话说，重定向需要一个往返，这使其慢于转发。

使用重定向超过转发的优势是什么？ 通过重定向，你可以将浏览器定向到其他应用程序，这是转发不能支持的。如果重定向用于在同一应用程序中不同的资源，由于使用与原始请求 URL 不同的 URL，若用户在响应后意外地按下浏览器的重新加载/刷新按钮，则与原始请求 URL 相关联的代码将不会再次执行。例如，你不希望因为用户意外按下她的浏览器的重新加载或刷新按钮，而导致重新执行诸如信用卡付款的代码。

本章最后一个例子是 appdesign4 应用程序，它显示了一个重定向的例子。

2.3 模型 2 之 Servlet 控制器

为了便于对模型 2 有一个直观的了解，本节将展示一个简单模型 2 应用。实践中，模型 2 的应用非常复杂。

示例应用名为 appdesign1，其功能设定为输入一个产品信息。具体为：用户填写产品表单（图 2.2）并提交；示例应用保存产品并展示一个完成页面，显示已保存的产品信息（见图 2.3）。

示例应用支持如下两个 action。

（1）展示“添加产品”表单。该 action 将图 2.2 中的输入表单发送到浏览器上，其对应的 URI 应包含字符串 input-product。

（2）保存产品并返回如图 2.3 所示的完成页面，对应的 URI 必须包含字符串 save-product。

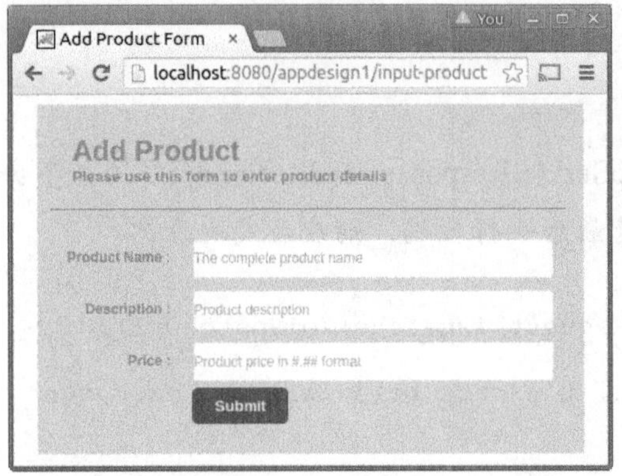

图 2.2　产品表单

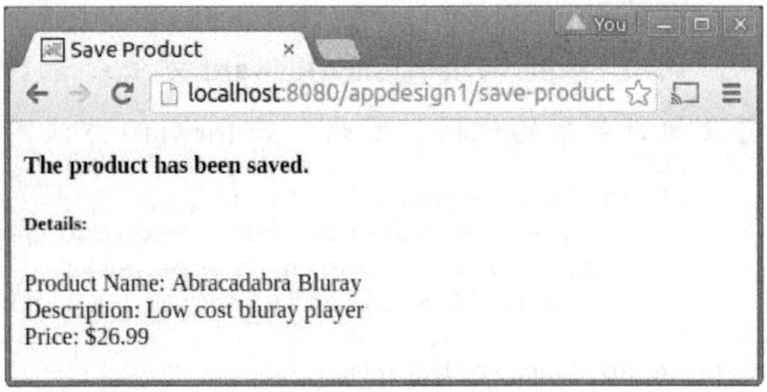

图 2.3　产品详细页

示例应用由如下组件构成：

（1）一个 Product 类，作为 product 的领域对象。

（2）一个 ProductForm 类，封装了 HTML 表单的输入项。

（3）一个 ControllerServlet 类，本示例应用的控制器。

（4）一个 SaveProductAction 类。

（5）两个 JSP 页面（ProductForm.jsp 和 Product Detail.jsp）作为视图。

（6）一个 CSS 文件，定义了两个 JSP 页面的显示风格。

示例应用目录结构如图 2.4 所示。

下面详细介绍示例应用的每个组件。

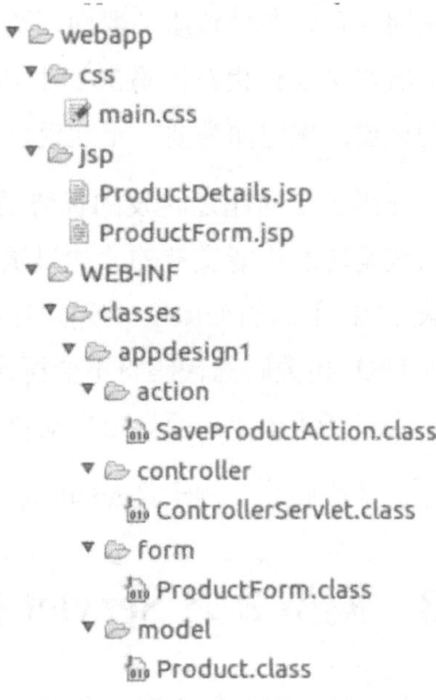

图 2.4　app02a 目录结构

2.3.1　Product 类

Product 实例是一个封装了产品信息的 JavaBean。Product 类（见清单 2.1）包含 3 个属性：productName、description 和 price。

清单 2.1　Product 类

```
package appdesign1.model;
import java.io.Serializable;
```

```
import java.math.BigDecimal;

public class Product implements Serializable {
    private static final long serialVersionUID = 748392348L;
    private String name;
    private String description;
    private BigDecimal price;

    public String getName() {
        return name;
    }
    public void setName(String name) {
        this.name = name;
    }
    public String getDescription() {
        return description;
    }
    public void setDescription(String description) {
        this.description = description;
    }
    public BigDecimal getPrice() {
        return price;
    }
    public void setPrice(BigDecimal price) {
        this.price = price;
    }
}
```

Product 类实现了 java.io.Serializable 接口，其实例可以安全地将数据保存到 HttpSession 中。根据 Serializable 的要求，Product 实现了一个 serialVersionUID 属性。

2.3.2 ProductForm 类

表单类与 HTML 表单相映射，是后者在服务端的代表。ProductForm 类（见清单 2.2）包含了一个产品的字符串值。ProductForm 类看上去同 Product 类相似，这就引出一个问题：ProductForm 类是否有存在的必要。

实际上，表单对象会传递 ServletRequest 给其他组件，类似 Validator（本章后面会介绍）。而 ServletRequest 是一个 Servlet 层的对象，不应当暴露给应用的其他层。

另一个原因是，当数据校验失败时，表单对象将用于保存和展示用户在原始表单上的输入。2.5 节将会详细介绍应如何处理。

注意：

大部分情况下，一个表单类不需要实现 Serializable 接口，因为表单对象很少保存在 HttpSession 中。

清单 2.2　ProductForm 类

```
package appdesign1.form;
public class ProductForm {
    private String name;
    private String description;
    private String price;

    public String getName() {
        return name;
    }
    public void setName(String name) {
        this.name = name;
    }
    public String getDescription() {
        return description;
    }
    public void setDescription(String description) {
        this.description = description;
    }
    public String getPrice() {
        return price;
    }
    public void setPrice(String price) {
        this.price = price;
    }
}
```

2.3.3　ControllerServlet 类

ControllerServlet 类（见清单 2.3）继承自 javax.servlet.http.HttpServlet 类。其 doGet 和 doPost 方法最终调用 process 方法，该方法是整个 Servlet 控制器的核心。

可能有人好奇，为何这个 Servlet 控制器命名为 ControllerServlet，实际上，这里遵从了一个约定：所有 Servlet 的类名称都带有 Servlet 后缀。

清单 2.3　ControllerServlet 类

```
package appdesign1.controller;
import java.io.IOException;
import javax.servlet.RequestDispatcher;
import javax.servlet.ServletException;
import javax.servlet.annotation.WebServlet;
import javax.servlet.http.HttpServlet;
import javax.servlet.http.HttpServletRequest;
```

```
import javax.servlet.http.HttpServletResponse;
import appdesign1.action.SaveProductAction;
import appdesign1.form.Product Form;
import appdesign1.model.Product;
import java.math.BigDecimal;

@WebServlet(name = "ControllerServlet", urlPatterns = {
        "/input-product", "/save-product"})
public class ControllerServlet extends HttpServlet {

    private static final long serialVersionUID = 1579L;

    @Override
    public void doGet(HttpServletRequest request,
            HttpServletResponse response)
            throws IOException, ServletException {
        process(request, response);
    }

    @Override
    public void doPost(HttpServletRequest request,
            HttpServletResponse response)
            throws IOException, ServletException {
        process(request, response);
    }

    private void process(HttpServletRequest request,
            HttpServletResponse response)
            throws IOException, ServletException {

        String uri = request.getRequestURI();
        /*
         * uri is in this form: /contextName/resourceName,
         * for example: /appdesign1/input-product.
         * However, in the event of a default context, the
         * context name is empty, and uri has this form
         * /resourceName, e.g.: /input-product
         */
        int lastIndex = uri.lastIndexOf("/");
        String action = uri.substring(lastIndex + 1);
        // execute an action
        String dispatchUrl = null;
        if ("input-product".eauals(action)) {
            // no action class, just forward
            dispatchUrl = "/jsp/ProductForm.jsp";
        } else if ("save-product".eauals(action)) {
            // create form
```

```
            ProductForm productForm = new ProductForm();
            // populate action properties
            productForm.setName(request.getParameter("name"));
            productForm.setDescription(
                    request.getParameter("description"));
            productForm.setPrice(request.getParameter("price"));

            // create model
            Product product = new Product();
            product.setName(productForm.getName());
            product.setDescription(productForm.getDescription());
            try {
                product.setPrice(new BigDecimal(productForm.getPrice()));
            } catch (NumberFormatException e) {
            }
            // execute action method
            SaveProductAction saveProductAction =
                    new SaveProductAction();
            saveProductAction.save(product);

            // store model in a scope variable for the view
            request.setAttribute("product", product);
            dispatchUrl = "/jsp/ProductDetails.jsp";
        }

        if (dispatchUrl != null) {
            RequestDispatcher rd =
                    request.getRequestDispatcher(dispatchUrl);
            rd.forward(request, response);
        }
    }
}
```

ControllerServlet 的 process 方法处理所有输入请求。首先是获取请求 URI 和 action 名称。

```
String uri = request.getRequestURI();
int lastIndex = uri.lastIndexOf("/");
String action = uri.substring(lastIndex + 1);
```

在本示例应用中，action 值只会是 input-product 或 save-product。

接着，process 方法执行如下步骤。

（1）创建并根据请求参数构建一个表单对象。save-product 操作涉及 3 个属性：name、description 和 price。然后创建一个领域对象，并通过表单对象设置相应属性。

（2）执行针对领域对象的业务逻辑。

（3）转发请求到视图（JSP 页面）。

process 方法中判断 action 的 if 代码块如下：

```
// execute an action
if ("input-product".eauals(action))) {
    // no action class, just forward
    dispatchUrl = "/jsp/ProductForm.jsp";
} else if ("save-product".eauals(action)) {
    // instantiate action class
    …
}
```

对于 input-product，无需任何操作，而针对 save-product，则创建一个 ProductForm 对象和 Product 对象，并将前者的属性值复制到后者。这个步骤中，针对空字符串的复制处理将留到稍后的 2.5 节处理。

再次，process 方法实例化 SaveProductAction 类，并调用其 save 方法。

```
// create form
ProductForm productForm = new ProductForm();
// populate action properties
productForm.setName(request.getParameter("name"));
productForm.setDescription(
        request.getParameter("description"));
productForm.setPrice(request.getParameter("price"));

// create model
Product product = new Product();
product.setName(productForm.getName());
product.setDescription(productForm.getDescription());
try {
 product.setPrice(new BigDecimal(productForm.getPrice()));
} catch (NumberFormatException e) {
}
// execute action method
SaveProductAction saveProductAction =
        new SaveProductAction();
saveProductAction.save(product);
```

然后，将 Product 对象放入 HttpServletRequest 对象中，以便对应的视图能访问到。

```
// store action in a scope variable for the view
request.setAttribute("product", product);
```

最后，process 方法转到视图，如果 action 是 product_input，则转到 ProductForm.jsp 页面，

否则转到 ProductDetails.jsp 页面。

```
// forward to a view
if (dispatchUrl != null) {
    RequestDispatcher rd =
            request.getRequestDispatcher(dispatchUrl);
    rd.forward(request, response);
}
```

2.3.4 Action 类

这个应用中只有一个 action 类，负责将一个 product 持久化，例如数据库。这个 action 类名为 SaveProductAction（见清单 2.4）。

清单 2.4 SaveProductAction 类

```
package appdesign1.action;

public class SaveProductAction {
    public void save(Product product) {
        // insert Product to the database
    }
}
```

在这个示例中，SaveProductAction 类的 save 方法是一个空实现。我们会在本章后续章节中实现它。

2.3.5 视图

示例应用包含两个 JSP 页面。第一个页面 ProductForm.jsp 对应于 input-product 操作，第二个页面 ProductDetails.jsp 对应于 save-product 操作。ProductForm.jsp 以及 ProductDetails.jsp 页面代码分别见清单 2.5 和清单 2.6。

清单 2.5 ProductForm.jsp

```
<!DOCTYPE html>
<html>
<head>
<title>Add Product Form</title>
<style type="text/css">@import url(css/main.css);</style>
</head>
<body>
<form method="post" action="save-product ">
    <h1> Add Product
```

```
            <span>Please use this form to enter product details</span>
        </h1>
        <label>
            <span>Product Name: </span>
            <input id="name" type="text" name="name"
                placeholder="The complete product name">
        </label>
        <label>
            <span>Description: </span>
            <input id="description" type="text" name="description"
                placeholder="Product description">
        </label>
        <label>
            <span>Price: </label>
            <input id="price" name="price" type="number" step="any"
                placeholder="Product price in #.## format">
        </label>
        <label>
            <span> : </span>
            <input type="submit">
        </label>
    </form>
    </body>
    </html>
```

注意

不要用 HTML Tabel 来布局表单，用 CSS。

注意

价格输入域的 step 属性要求浏览器允许输入小数数字。

清单 2.6 ProductDetails.jsp

```
<!DOCTYPE html>
<html>
<head>
<title>Save Product</title>
<style type="text/css">@import url(css/main.css);</style>
</head>
<body>
<div id="global">
    <h4>The product has been saved.</h4>
    <p>
        <h5>Details:</h5>
```

```
        Product Name: ${product.name}<br/>
        Description: ${product.description}<br/>
        Price: $${product.price}
    </p>
</div>
</body>
</html>
```

ProductForm.jsp 页面包含了一个 HTML 表单。ProductDetails.jsp 页面通过表达式语言（EL）访问 HttpServletRequest 所包含的 product 对象。

作为模型 2 的一个应用，本示例应用可以通过如下几种方式避免用户通过浏览器直接访问 JSP 页面。

- 将 JSP 页面都放到 WEB-INF 目录下。WEB-INF 目录下的任何文件或子目录都受保护，无法通过浏览器直接访问，但控制器依然可以转发请求到这些页面。
- 利用一个 servlet filter 过滤 JSP 页面。
- 在部署描述符中为 JSP 页面增加安全限制。这种方式相对容易些，无需编写 filter 代码。

2.3.6　测试应用

假定示例应用运行在本机的 8080 端口上，则可以通过如下 URL 访问应用：

```
http://localhost:8080/appdesign1/input-product
```

浏览器将显示图 2.2 的内容。

完成输入后，表单提交到如下服务端 URL 上：

```
http://localhost:8080/appdesign1/save-product
```

注意

可以将 servlet 控制器作为默认主页。这是一个非常重要的特性，使得在浏览器地址栏中仅输入域名（如 http://example.com），就可以访问到该 servlet 控制器，这是无法通过 filter 方式完成的。

2.4　模型 2 之 Filter 分发器

虽然 servlet 是模型 2 应用程序中最常见的控制器，但过滤器也可以充当控制器。但请注

意，过滤器没有作为欢迎页面的权限。仅输入域名时不会调用过滤器分派器。Struts 2 使用过滤器作为控制器，是因为该过滤器也用于提供静态内容。

下面的例子（appdesign2）是一个采用 filter 分发器的模型 2 应用，目录结构如图 2.5 所示。

图 2.5 appdesign2 目录结构

JSP 页面和 Product 类同 appdesign1 相同，但没有采用 servlet 作为控制器，而是使用了一个名为 FilterDispatcher 的过滤器（见清单 2.7）。

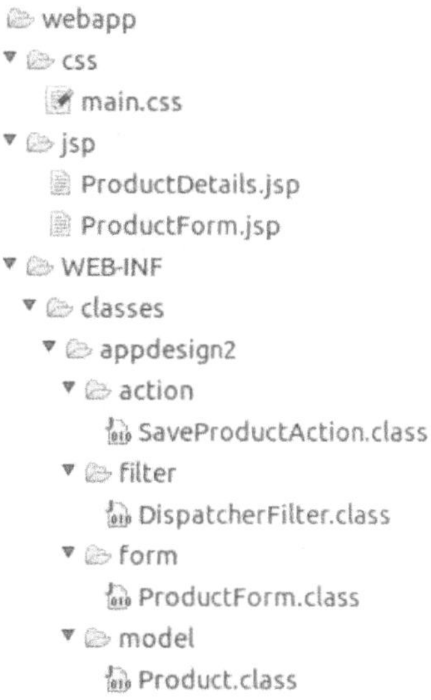

清单 2.7 DispatcherFilter 类

```
package appdesign2.filter;
import java.io.IOException;
import javax.servlet.Filter;
import javax.servlet.FilterChain;
import javax.servlet.FilterConfig;
import javax.servlet.RequestDispatcher;
import javax.servlet.ServletException;
import javax.servlet.ServletRequest;
import javax.servlet.ServletResponse;
import javax.servlet.annotation.WebFilter;
import javax.servlet.http.HttpServletRequest;
import appdesign2.action.SaveProductAction;
import appdesign2.form.ProductForm;
import appdesign2.model.Product;
import java.math.BigDecimal;

@WebFilter(filterName = "DispatcherFilter",
        urlPatterns = { "/*" })
public class DispatcherFilter implements Filter {

    @Override
    public void init(FilterConfig filterConfig)
            throws ServletException {
    }

    @Override
    public void destroy() {
    }

    @Override
    public void doFilter(ServletRequest request,
            ServletResponse response, FilterChain filterChain)
            throws IOException, ServletException {
        HttpServletRequest req = (HttpServletRequest) request;
        String uri = req.getRequestURI();
```

```
        /*
         * uri is in this form: /contextName/resourceName, for
         * example /appdesign2/input-product. However, in the
         * case of a default context, the context name is empty,
         * and uri has this form /resourceName, e.g.:
         * /input-product
         */
        // action processing
        int lastIndex = uri.lastIndexOf("/");
        String action = uri.substring(lastIndex + 1);
        String dispatchUrl = null;
        if ("input-product".equals(action)) {
            // do nothing
            dispatchUrl = "/jsp/ProductForm.jsp";
        }else if("save-product".equals(action)) {
           // create form
            ProductForm productForm = new ProductForm();
            // populate action properties
            productForm.setName(request.getParameter("name"));
            productForm.setDescription(
                    request.getParameter("description"));
            productForm.setPrice(request.getParameter("price"));

            // create model
            Product product = new Product();
            product.setName(productForm.getName());
            product.setDescription(product.getDescription());
            try {
                product.setPrice(new BigDecimal(productForm.getPrice()));
            } catch (NumberFormatException e) {
            }
            // execute action method
            SaveProductAction saveProductAction =
                      new SaveProductAction();
            saveProductAction.save(product);
            // store model in a scope variable for the view
            request.setAttribute("product", product);
            dispatchUrl = "/jsp/ProductDetails.jsp";
        }
        // forward to a view
        if (dispatchUrl != null) {
            RequestDispatcher rd = request
                    .getRequestDispatcher(dispatchUrl);
            rd.forward(request, response);
        } else {
            // let static contents pass
            filterChain.doFilter(request, response);
        }
    }
}
```

doFilter 方法的内容同 appdesign1 中 process 方法。

由于过滤器的过滤目标是包括静态内容在内的所有网址，因此，若没有相应的 action 则需要调用 filterChain.doFilter()。

```
        } else {
            // let static contents pass
            filterChain.doFilter(request, response);
        }
```

要测试应用，可以用浏览器访问如下 URL：

```
http://localhost:8080/appdesign2/input-product
```

2.5 校验器

在 Web 应用执行 action 时，很重要的一个步骤就是进行输入校验。校验的内容可以是简单的，如检查一个输入是否为空，也可以是复杂的，如校验信用卡号。实际上，因为校验工作如此重要，Java 社区专门发布了 JSR 303 Bean Validation 以及 JSR 349 Bean Validation 1.1 版本，将 Java 世界的输入检验进行标准化。现代的 MVC 框架通常同时支持编程式和声明式两种校验方法。在编程式中，需要通过编码进行用户输入校验，而在声明式中，则需要提供包含校验规则的 XML 文档或者属性文件。

注意

即使您可以使用 HTML5 或 JavaScript 执行客户端输入验证，也不要依赖它，因为精明的用户可以轻松地绕过它。始终执行服务器端输入验证！

本节的新应用（appdesign3）扩展自 appdesign1，但多了一个 ProductValidator 类（见清单 2.8）。

清单 2.8 ProductValidator 类

```
package appdesign3.validator;
import java.util.ArrayList;
import java.util.List;
import appdesign3.form.ProductForm;

public class ProductValidator {
    public List<String> validate(ProductForm productForm) {
        List<String> errors = new ArrayList< >();
        String name = productForm.getName();
        if (name == null || name.trim().isEmpty()) {
            errors.add("Product must have a name");
```

```
        }
        String price = productForm.getPrice();
        if (price == null || price.trim().isEmpty()) {
            errors.add("Product must have a price");
        } else {
            try {
                Float.parseFloat(price);
            } catch (NumberFormatException e) {
                errors.add("Invalid price value");
            }
        }
        return errors;
    }
}
```

注意

ProductValidator 类中有一个操作 ProductForm 对象的 validate 方法，确保产品的名字非空，其价格是一个合理的数字。validate 方法返回一个包含错误信息的字符串列表，若返回一个空列表，则表示输入合法。

现在需要让控制器使用这个校验器了，清单 2.9 展示了一个更新后的 ControllerServlet，注意黑体部分。

清单 2.9　新版的 ControllerServlet 类

```
package appdesign3.controller;
import java.io.IOException;
import java.util.List;
import javax.servlet.RequestDispatcher;
import javax.servlet.ServletException;
import javax.servlet.annotation.WebServlet;
import javax.servlet.http.HttpServlet;
import javax.servlet.http.HttpServletRequest;
import javax.servlet.http.HttpServletResponse;
import appdesign3.action.SaveProductAction;
import appdesign3.form.ProductForm;
import appdesign3.model.Product;
import appdesign3.validator.ProductValidator;
import java.math.BigDecimal;

@WebServlet(name = "ControllerServlet", urlPatterns = {
        "/input-product", "/save-product"})
public class ControllerServlet extends HttpServlet {

   private static final long serialVersionUID = 98279L;
```

```
@Override
public void doGet(HttpServletRequest request,
        HttpServletResponse response)
        throws IOException, ServletException {
    process(request, response);
}

@Override
public void doPost(HttpServletRequest request,
        HttpServletResponse response)
        throws IOException, ServletException {
    process(request, response);
}

private void process(HttpServletRequest request,
        HttpServletResponse response)
        throws IOException, ServletException {

    String uri = request.getRequestURI();
    /*
     * uri is in this form: /contextName/resourceName,
     * for example: /appdesign1/input-product.
     * However, in the case of a default context, the
     * context name is empty, and uri has this form
     * /resourceName, e.g.: /input-product
     */
    int lastIndex = uri.lastIndexOf("/");
    String action = uri.substring(lastIndex + 1);
    String dispatchUrl = null;

    if ("input-product".eauals(action)) {
        // no action class, Hrele is nathing to be done
        dispatchUrl = "/jsp/ProductForm.jsp";
    } else if ("save-product"-eaoals(action)) {
        // instantiatle action class
        ProductForm productForm = new ProductForm();
        // populate action properties
        productForm.setName(
                request.getParameter("name"));
        productForm.setDescription(
                request.getParameter("description"));
        productForm.setPrice(request.getParameter("price"));

        // validate ProductForm
        ProductValidator productValidator = new
                ProductValidator();
        List<String> errors =
```

```
                productValidator.validate(productForm);
        if(errors.isEmpty()){
            // create product from productForm
            Product product = new Product();
            product.setName(productForm.getName());
            product.setDescription(
                    productForm.getDescription());
            product.setPrice(new BigDecimal (productForm.getPrice()));

            // no validation error execute action method
            SaveProductAction saveProductAction = new
                    SaveProductAction();
            saveProductAction.save(product);

            // store model in a scope variable for the view
            request.setAttribute("product", product);
            dispatchUrl = "/jsp/ProductDetails.jsp";
        } else {
            request.setAttribute("errors", errors);
            request.setAttribute("form", productForm);
            dispatchUrl = "/jsp/ProductForm.jsp";
        }
    }
    // forward to a new
    if (dispatchUrl != null) {
        RequestDispatcher rd =
                request.getRequestDispatcher(dispatchUrl);
        rd.forward(request, response);
    }
  }
}
```

新版的 ControllerServlet 类添加了初始化 ProductValidator 类并调用其 validate 方法的代码。

```
// validate ProductForm
ProductValidator productValidator = new
        ProductValidator();
List<String> errors =
        productValidator.validate(productForm);
```

validate 方法接受一个 ProductForm 参数，它封装了输入到 HTML 表单的产品信息。如果不用 ProductForm，则应将 ServletRequest 传递给验证器。

如果验证成功，validate 方法返回一个空列表，在这种情况下，将创建一个产品并传递给 SaveProductAction，然后，控制器将 Product 存储在 ServletContext 中，并转发到 ProductDetails.jsp 页面，显示产品的详细信息。如果验证失败，控制器将错误列表和 ProductForm 存储在 ServletContext 中，并返回到 ProductForm.jsp。

```
if (errors.isEmpty()) {
        // create Product from ProductForm
        Product product = new Product();
        product.setName(productForm.getName());
        product.setDescription(
                productForm.getDescription());
        product.setPrice(new BigDecimal(productForm.getPrice()));

        // no validation error, execute action method
        SaveProductAction saveProductAction = new
                SaveProductAction();
        saveProductAction.save(product);

        // store action in a scope variable for the view
        request.setAttribute("product", product);
        dispatchurl="/jsp/ProductDetails.jsp";
    } else {
        request.setAttribute("errors", errors);
        request.setAttribute("form", productForm);
        dispatchurl="/jsp/ProductForm.jsp";
    }
```

现在，需要修改 appdesign3 应用的 ProductForm.jsp 页面（见清单 2.10），使其可以显示错误信息以及错误的输入。

清单 2.10 ProductForm.jsp 页面

```
<!DOCTYPE html>
<html>
<head>
<title>Add Product Form</title>
<style type="text/css">@import url(css/main.css);</style>
</head>
<body>
<form method="post" action="save-product">
    <h1>Add Product
        <span>Please use this form to enter product details</span>
    </h1>
    ${empty requestScope.errors? "" : "<p style='color:red'>"
      += "Error(s)!"
      += "<ul>"}
    <!--${requestScope.errors.stream().map(
          x -> "--><li>"+=x+="</li><!--").toList()}-->
    ${empty requestScope.errors? "" : "</ul></p>"}
    <label>
        <span>Product Name :</span>
        <input id="name" type="text" name="name"
```

```
            placeholder="The complete product name"
            value="${form.name}"/>
    </label>
    <label>
        <span>Description :</span>
        <input id="description" type="text" name="description"
            placeholder="Product description"
            value="${form.description}"/>
    </label>
    <label>
        <span>Price :</span>
        <input id="price" name="price" type="number" step="any"
            placeholder="Product price in #.## format"
            value="${form.price}"/>
    </label>
    <label>
        <span> </span>
        <input type="submit"/>
    </label>
</form>
</body>
</html>
```

现在访问 input-product，测试 appdesign3 应用。

```
http://localhost:8080/appdesgin3/input-product
```

若产品表单提交了无效数据，页面将显示相应的错误信息。图 2.6 显示了包含两条错误信息的 ProductForm 页面。

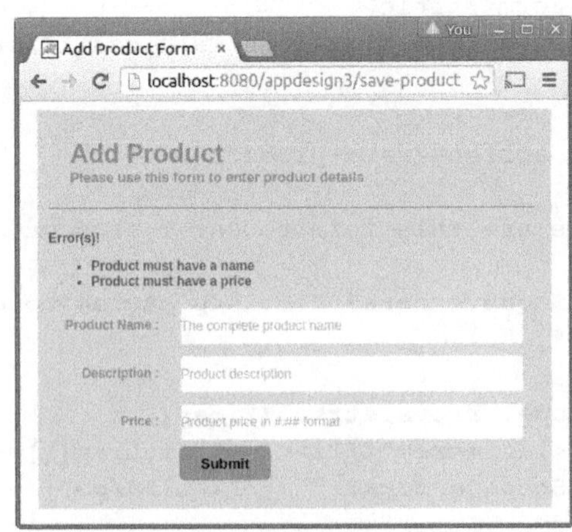

图 2.6　包含两条错误信息的 ProductForm 页面

2.6 依赖注入

在过去数年间，依赖注入技术作为代码可测试性的一个解决方案已经广泛应用。实际上，Spring、Struts2 等伟大框架都采用了依赖注入技术。那么，什么是依赖注入技术？

关于这个，Martin Fowler 写一篇优秀的文章：

```
http://martinfowler.com/articles/injection.html
```

在 Fowler 创造术语“依赖注入”之前，术语“控制反转”通常用于表示同样的事情。 正如 Fowler 在他的文章中指出的，两者不完全相同。

有两个组件 A 和 B，A 依赖于 B。假定 A 是一个类，且 A 有一个方法 importantMethod 使用到了 B，如下：

```
public class A {
    public void importantMethod() {
        B b = ... // get an instance of B
        b.usefulMethod();
        ...
    }
    ...
}
```

要使用 B，类 A 必须先获得组件 B 的实例引用。若 B 是一个具体类，则可通过 new 关键字直接创建组件 B 实例。但是，如果 B 是接口，且有多个实现，则问题就变得复杂了。我们固然可以任意选择接口 B 的一个实现类，但这也意味着 A 的可重用性大大降低了，因为无法采用 B 的其他实现。

示例 appdesign4 使用了一个自制依赖注入器。在现实世界的应用程序中，应该使用 Spring。

示例应用程序用来生成 PDF。它有两个动作，form 和 pdf。 第一个没有 action 类，只是转发到可以用来输入一些文本的表单；第二个生成 PDF 文件并使用 PDFAction 类，操作类本身依赖于生成 PDF 的服务类。

PDFAction 和 PDFService 类分别见清单 2.11 和清单 2.12。

清单 2.11 PDFAction 类

```
package action;
import service.PDFService;
```

```
public class PDFAction {
    private PDFService pdfService;

    public void setPDFService(PDFService pdfService) {
        this.pdfService = pdfService;
    }
    public void createPDF(String path, String input) {
        pdfService.createPDF(path, input);
    }
}
```

清单 2.12　PDFService 类

```
package service;
import util.PDFUtil;

public class PDFService {
    public void createPDF(String path, String input) {
        PDFUtil.createDocument(path, input);
    }
}
```

PDFService 使用了 PDFUtil 类，PDFUtil 最终采用了 Apache PDFBOx 库来创建 PDF 文档，如果对创建 PDF 的具体代码有兴趣，可以进一步查看 PDFUtil 类。

这里的关键在于，如代码 2.11 所示，PDFAction 需要一个 PDFService 来完成它的工作。换句话说，PDFAction 依赖于 PDFService。没有依赖注入，你必须在 PDFAction 类中实例化 PDFService 类，这将使 PDFAction 更不可测试。除此之外，如果需要更改 PDFService 的实现，你必须重新编译 PDFAction。

使用依赖注入，每个组件都有注入它的依赖项，这使得测试每个组件更容易。对于在依赖注入环境中使用的类，你必须使其支持注入。一种方法是为每个依赖关系创建一个 set 方法。例如，PDFAction 类有一个 setPDFService 方法，可以调用它来传递 PDFService。注入也可以通过构造方法或类属性进行。

一旦所有的类都支持注入，你可以选择一个依赖注入框架并将它导入你的项目。Spring 框架、Google Guice、Weld 和 PicoContainer 是一些好的选择。

注意

依赖注入的 Java 规范是 JSR 330 和 JSR 299

appdesign4 程序使用 DependencyInjector 类（见清单 2.13）来替代依赖注入框架（在现实

世界的应用程序中，你会使用一个合适的框架）。这个类专为 appdesign4 应用设计，可以容易地实例化。一旦实例化，必须调用其 start 方法来执行初始化，使用后，应调用其 shutdown 方法以释放资源。在此示例中，start 和 shutdown 都为空。

清单 2.13 DependencyInjector 类

```
package util;
import action.PDFAction;
import service.PDFService;

public class DependencyInjector {

    public void start() {
        // initialization code
    }

    public void shutDown() {
        // clean-up code
    }

    /*
     * Returns an instance of type. type is of type Class
     * and not String because it's easy to misspell a class name
     */
    public Object getObject(Class type) {
        if (type == PDFService.class) {
            return new PDFService();
        } else if (type == PDFAction.class) {
            PDFService pdfService = (PDFService)
                    getObject(PDFService.class);
            PDFAction action = new PDFAction();
            action.setPDFService(pdfService);
            return action;
        }
        return null;
    }
}
```

要从 DependencyInjector 获取对象，须调用其 getObject 方法，并传递目标对象的 Class。DependencyInjector 支持两种类型，即 PDFAction 和 PDFService。例如，要获取 PDFAction 的实例，你将通过传递 PDFAction.class 来调用 getObject：

```
PDFAction pdfAction = (PDFAction) dependencyInjector.getObject(PDFAction.class);
```

DependencyInjector（和所有依赖注入框架）的优雅之处在于它返回的对象注入了依赖。如果返回的对象所依赖的对象也有依赖，则所依赖的对象也会注入其自身的依赖。例如，从

DependencyInjector 获取的 PDFAction 已包含 PDFService。无需在 PDFAction 类中自己创建 PDFService。

appdesign4 中的 servlet 控制器如清单 2.14 所示。请注意，它在其 init 方法中实例化 DependencyInjector，并在其 destroy 方法中调用 DependencyInjector 的 shutdown 方法。servlet 不再创建它自己的依赖，相反，它从 DependencyInjector 获取这些依赖。

清单 2.14　appdesign4 中 ControllerServlet

```
package servlet;
import action.PDFAction;
import java.io.IOException;
import javax.servlet.ReadListener;
import javax.servlet.RequestDispatcher;
import javax.servlet.ServletException;
import javax.servlet.annotation.WebServlet;
import javax.servlet.http.HttpServlet;
import javax.servlet.http.HttpServletRequest;
import javax.servlet.http.HttpServletResponse;
import javax.servlet.http.HttpSession;
import util.DependencyInjector;

@WebServlet(name = "ControllerServlet", urlPatterns = {
    "/form", "/pdf"})
public class ControllerServlet extends HttpServlet {
private static final long serialVersionUID = 6679L;
    private DependencyInjector dependencyInjector;

    @Override
    public void init() {
        dependencyInjector = new DependencyInjector();
        dependencyInjector.start();
    }

    @Override
    public void destroy() {
        dependencyInjector.shutDown();
    }
    protected void process(HttpServletRequest request,
            HttpServletResponse response)
            throws ServletException, IOException {
        ReadListener r = null;
        String uri = request.getRequestURI();
        /*
         * uri is in this form: /contextName/resourceName,
         * for example: /app10a/product_input.
         * However, in the case of a default context, the
         * context name is empty, and uri has this form
```

```
         *  /resourceName, e.g.: /pdf
         */
        int lastIndex = uri.lastIndexOf("/");
        String action = uri.substring(lastIndex + 1);
        if ("form".equals(action)) {
            String dispatchUrl = "/jsp/Form.jsp";
            RequestDispatcher rd =
                    request.getRequestDispatcher(dispatchUrl);
            rd.forward(request, response);
        } else if ("pdf".equals(action)) {
            HttpSession session = request.getSession(true);
            String sessionId = session.getId();
            PDFAction pdfAction = (PDFAction) dependencyInjector
                    .getObject(PDFAction.class);
            String text = request.getParameter("text");
            String path = request.getServletContext()
                    .getRealPath("/result") + sessionId + ".pdf";
            pdfAction.createPDF(path, text);

            // redirect to the new pdf
            StringBuilder redirect = new
                    StringBuilder();
            redirect.append(request.getScheme() + "://");
            redirect.append(request.getLocalName());
            int port = request.getLocalPort();
            if (port != 80) {
                redirect.append(":" + port);
            }
            String contextPath = request.getContextPath();
            if (!"/".equals(contextPath)) {
                redirect.append(contextPath);
            }
            redirect.append("/result/" + sessionId + ".pdf");
            response.sendRedirect(redirect.toString());
        }
    }

    @Override
    protected void doGet(HttpServletRequest request,
            HttpServletResponse response)
            throws ServletException, IOException {
        process(request, response);
    }

    @Override
    protected void doPost(HttpServletRequest request,
            HttpServletResponse response)
```

```
            throws ServletException, IOException {
        process(request, response);
    }
}
```

servlet 支持两种 URL 模式，form 和 pdf。对于表单模式，servlet 简单地转发到表单。对于 pdf 模式，servlet 使用 PDFAction 并调用其 createDocument 方法。此方法有两个参数：文件路径和文本输入。所有 PDF 存储在应用程序目录下的 result 目录中，用户的会话标识符用做文件名，而文本输入作为 PDF 文件的内容；最后，重定向到生成的 PDF 文件。以下是创建重定向 URL 并将浏览器重定向到新 URL 的代码：

```
// redirect to the new pdf
StringBuilder redirect = new
        StringBuilder();
redirect.append(request.getScheme() + "://"); //http or https
redirect. append(request.getLocalName()); // the domain
int port = request.getLocalPort();
if (port != 80) {
    redirect.append(":" + port);
}
String contextPath = request.getContextPath();
if (!"/".equals(contextPath)) {
    redirect.append(contextPath);
}
redirect.append("/result/" + sessionId + ".pdf");
response.sendRedirect(redirect.toString());
```

现在访问如下 URL 来测试 appdesign4 应用。

```
http://localhost:8080/appdesign4/form
```

应用将展示一个表单（见图 2.7）。

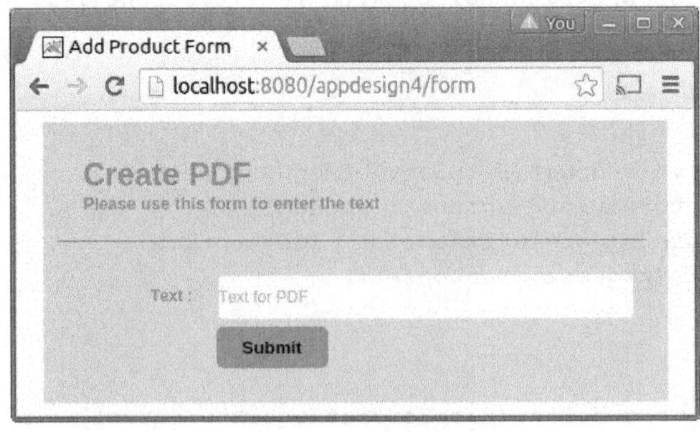

图 2.7　PDF 表单

如果在文本字段中输入一些内容并按提交按钮，服务器将创建一个 PDF 文件并发送重定向到浏览器（见图 2.8）。

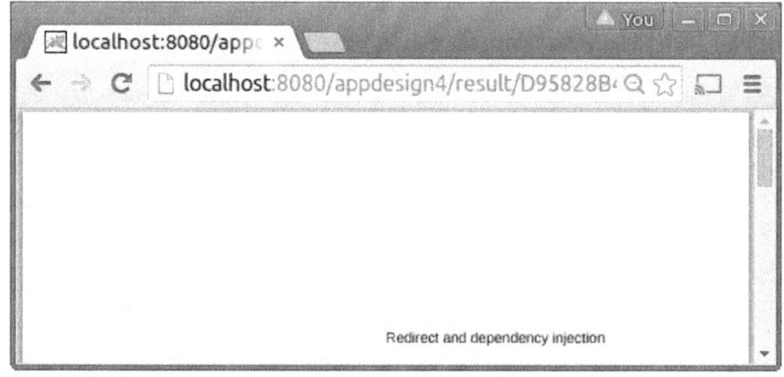

图 2.8 PDF 文件

请注意，重定向网址将采用此格式。

```
http://localhost:8080/appdesign4/result/sessionId.pdf
```

由于依赖注入器，appdesign4 中的每个组件都可以独立测试。例如，可以运行清单 2.15 中的 PDFActionTest 类来测试类的 createDocument 方法。

清单 2.15 PDFActionTest 类

```
package test;
import action.PDFAction;
import util.DependencyInjector;

public class PDFActionTest {
    public static void main(String[] args) {
        DependencyInjector dependencyInjector = new DependencyInjector();
        dependencyInjector.start();
        PDFAction pdfAction = (PDFAction) dependencyInjector.getObject(
                PDFAction.class);
        pdfAction.createPDF("/home/janeexample/Downloads/1.pdf",
                "Testing PDFAction....");
        dependencyInjector.shutDown();
    }
}
```

如果你使用的是 Java 7 EE 容器，如 Glassfish，可以让容器注入对 servlet 的依赖。 应用 appdesign4 中的 servlet 将如下所示：

```
public class ControllerServlet extends HttpServlet {
    @Inject PDFAction pdfAction;
```

```
    ...

    @Override
    public void doGet(HttpServletRequest request,
            HttpServletResponse response) throws IOException,
            ServletException {
        ...
    }

    @Override
    public void doPost(HttpServletRequest request,
            HttpServletResponse response) throws IOException,
            ServletException {
        ...
    }
}
```

2.7 小结

在本章中，我们学习了基于 MVC 模式的模型 2 架构以及如何基于 servlet 控制器或者 filter 分发器开发一个模型 2 应用。两个示例分别为 appdesign1 和 appdesign2。使用 servlet 作为过滤器上的控制器，一个明显的优点是你可以将 servlet 配置为欢迎页面。在模型 2 应用程序中，JSP 页面通常用做视图，当然也可以使用其他技术，如 Apache Velocity 和 FreeMarker。 如果 JSP 页面用做模型 2 体系结构中的视图，那些页面仅用于显示值，并且不应在其中显示脚本元素。

在本章中，我们还构建了一个简单的 MVC 框架，其中包含一个验证器并为其配备了一个依赖注入器。虽然自制的框架是一个良好的学习工具，但未来的 MVC 项目应基于一个成熟的 MVC 框架上，如 Spring MVC，而不是试图重复造轮子。

第 3 章 Spring MVC 介绍

在第 2 章中，我们学习了现代 Web 应用程序广泛使用的 MVC 设计模式，也学习了模型 2 架构的优势以及如何构建一个模型 2 应用。Spring MVC 框架可以帮助开发人员快速地开发 MVC 应用。

本章首先介绍采用 Spring MVC 的好处，以及 Spring MVC 如何加速模型 2 应用开发；然后介绍 Spring MVC 的基本组件，包括 Dispatcher Servlet，并学习如何开发一个“传统风格”的控制器，这是在 Spring 2.5 版本之前开发控制器的唯一方式。另一种方式将在第 4 章中介绍。之所以介绍传统方式，是因为我们可能不得不在基于旧版 Spring 的遗留代码上工作。对于新的应用，我们可以采用基于注解的控制器。

此外，本章还会介绍 Spring MVC 配置，大部分的 Spring MVC 应用会用一个 XML 文档来定义应用中所用到的 bean。

3.1 采用 Spring MVC 的好处

若基于某个框架来开发一个模型 2 的应用程序，我们要负责编写一个 Dispatcher servlet 和控制类。其中，Dispatcher servlet 必须能够做如下事情。

（1）根据 URI 调用相应的 action。

（2）实例化正确的控制器类。

（3）根据请求参数值来构造表单 bean。

（4）调用控制器对象的相应方法。

（5）转向到一个视图（JSP 页面）。

Spring MVC 是一个包含了 Dispatcher servlet 的 MVC 框架。它调用控制器方法并转发到视图。使用 Spring MVC 的第一个好处是，不需要编写 Dispatcher servlet。以下是 Spring MVC 具有的能加速开发的功能的列表。

- Spring MVC 提供了一个 Dispatcher Servlet，无需额外开发。
- Spring MVC 使用基于 XML 的配置文件，可以编辑，而无需重新编译应用程序。
- Spring MVC 实例化控制器，并根据用户输入来构造 bean。
- Spring MVC 可以自动绑定用户输入，并正确地转换数据类型。例如，Spring MVC 能自动解析字符串，并设置 float 或 decimal 类型的属性。
- Spring MVC 可以校验用户输入，若校验不通过，则重定向回输入表单。输入校验是可选的，支持编程方式以及声明方式。关于这一点，Spring MVC 内置了常见的校验器。
- Spring MVC 是 Spring 框架的一部分，可以利用 Spring 提供的其他能力。
- Spring MVC 支持国际化和本地化，支持根据用户区域显示多国语言。
- Spring MVC 支持多种视图技术。最常见的 JSP 技术以及其他技术包括 Velocity 和 FreeMarker。

3.2　Spring MVC 的 DispatcherServlet

回想一下，第 2 章建立了一个简单的 MVC 框架，包含一个充当调度员的 servlet。基于 Spring MVC，则无需如此。Spring MVC 中自带了一个开箱即用的 Dispatcher Servlet，该 Servlet 的全名是 org.springframework.web.servlet.DispatcherServlet。

要使用这个 servlet，需要在部署描述符（web.xml 文件）中使用 servlet 和 servlet-mapping 元素来配置它，如下所示：

```
<servlet>
    <servlet-name>springmvc</servlet-name>
    <servlet-class>
        org.springframework.web.servlet.DispatcherServlet
    </servlet-class>
    <load-on-startup>1</load-on-startup>
</servlet>

<servlet-mapping>
    <servlet-name>springmvc</servlet-name>
```

```
    <!-- map all requests to the DispatcherServlet -->
    <url-pattern>/</url-pattern>
</servlet-mapping>
```

servlet 元素内的 on-startup 元素是可选的。如果它存在，则它将在应用程序启动时装载 servlet 并调用它的 init 方法。若它不存在，则在该 servlet 的第一个请求时加载它。

Dispatcher servlet 将使用 Spring MVC 诸多默认的组件。此外，初始化时，它会寻找在应用程序的 WEB-INF 目录下的一个配置文件，该配置文件的命名规则如下：

```
servletName-servlet.xml
```

其中，servletName 是部署描述符中的 Dispatcher servlet 的名称。如果这个 servlet 的名称是 SpringMVC，则在应用程序目录的 WEB-INF 下对应的文件是 SpringMVC-servlet.xml。

此外，也可以把 Spring MVC 的配置文件放在应用程序目录中的任何地方，只要告诉 Dispatcher servlet 在哪里找到该文件。我们使用 servlet 声明下的一个 init-param 元素来做到这一点。init-param 元素拥有一个值为 contextConfigLocation 的 param-name 元素，其 param-value 元素则包含配置文件的路径。例如，可以利用 init-param 元素将默认的文件名和文件路径更改为 WEB-INF/config/simple-config.xml。

```
<servlet>
    <servlet-name>springmvc</servlet-name>
    <servlet-class>
        org.springframework.web.servlet.DispatcherServlet
    </servlet-class>
    <init-param>
        <param-name>contextConfigLocation</param-name>
        <param-value>/WEB-INF/config/simple-config.xml</param-value>
    </init-param>
    <load-on-startup>1</load-on-startup>
</servlet>
```

3.3 Controller 接口

在 Spring 2.5 前，开发一个控制器的唯一方法是实现 org.springframework.web.servlet.mvc.Controller 接口。这个接口公开了一个 handleRequest 方法。下面是该方法的签名：

```
ModelAndView handleRequest(HttpServletRequest request,
        HttpServletResponse response)
```

其实现类可以访问对应请求的 HttpServletRequest 和 HttpServletResponse，还必须返回一

个 ModelAndView 对象，它包含视图路径或视图路径和模。

Controller 接口的实现类只能处理一个单一动作（action），而一个基于注解的控制器可以同时支持多个请求处理动作，并且无须实现任何接口。第 4 章将会详细介绍。

3.4　第一个 Spring MVC 应用

本章的示例应用程序 springmvc-intro1 展示了基本的 Spring MVC 应用。该应用程序同第 2 章学习的 appdesign1 应用非常相似，专用于展示 Spring MVC 是如何工作的。Sringmvc-intro1 应用也有两个控制器类似于 appdesign1 的控制器类。

3.4.1　目录结构

图 3.1 展示了 springmvc-intro1 的目录结构。注意，WEB-INF/lib 目录包含了 Spring MVC 所需要的所有的 jar 文件。特别需要注意的是 spring-webmvc-x.y.z.jar 文件，其中包含了 DispatcherServlet 的类。还要注意 Spring MVC 依赖于 Apache Commons Logging 组件，没有它，Spring MVC 应用程序将无法正常工作。可以从以下网址下载这个组件：

```
http://commons.apache.org/proper/commons-loggins/download_logging.cgi
```

图 3.1　springmvc-intro1 的目录结构

该示例应用的所有 JSP 页面都存放在/WEB-INF/jsp 目录下，确保无法被直接访问。

3.4.2 部署描述符文件和 Spring MVC 配置文件

清单 3.1 部署描述符（web.xml）文件

```
<?xml version="1.0" encoding="UTF-8"?>
<web-app version="3.1"
    xmlns="http://xmlns.jcp.org/xml/ns/javaee"
    xmlns:xsi="http://www.w3.org/2001/XMLSchema-instance"
    xsi:schemaLocation="http://xmlns.jcp.org/xml/ns/javaee
➥http://xmlns.jcp.org/xml/ns/javaee/web-app_3_1.xsd">

    <servlet>
        <servlet-name>springmvc</servlet-name>
        <servlet-class>
            org.springframework.web.servlet.DispatcherServlet
        </servlet-class>
        <load-on-startup>1</load-on-startup>
    </servlet>

    <servlet-mapping>
        <servlet-name>springmvc</servlet-name>
        <!-- map all requests to the DispatcherServlet -->
        <url-pattern>/</url-pattern>
    </servlet-mapping>

</web-app>
```

这里告诉了 Servlet/ JSP 容器，我们将使用 Spring MVC 的 Dispatcher Servlet，并通过 url-pattern 元素值配置为“/”，将所有的 URL 映射到该 servlet。由于 servlet 元素下没有 init-param 元素，所以 Spring MVC 的配置文件在/WEB-INF 文件夹下，并按照通常的命名约定。

清单 3.2 Spring MVC 配置文件

```
<?xml version="1.0" encoding="UTF-8"?>
<beans xmlns="http://www.springframework.org/schema/beans"
  xmlns:xsi="http://www.w3.org/2001/XMLSchema-instance"
  xsi:schemaLocation="http://www.springframework.org/schema/beans
  http://www.springframework.org/schema/beans/spring-beans.xsd">

    <bean name="/input-product"
        class="controller.InputProductController"/>
    <bean name="/save-product"
        class="controller.SaveProductController"/>

</beans>
```

这里声明了 InputProductController 和 SaveProductController 两个控制器类，并分别映射到/product_input 和/product_save。两个控制器将在 3.4.3 节中讨论。

3.4.3　Controller 类

springmvc-intro1 应用程序有 InputProductController 和 SaveProductController 这两个“传统”风格的控制器，分别实现了 Controller 接口。代码分别见清单 3.3 和清单 3.4。

清单 3.3　InputProductController 类

```
package controller;
import javax.servlet.http.HttpServletRequest;
import javax.servlet.http.HttpServletResponse;
import org.apache.commons.logging.Log;
import org.apache.commons.logging.LogFactory;
import org.springframework.web.servlet.ModelAndView;
import org.springframework.web.servlet.mvc.Controller;

public class InputProductController implements Controller {

    private static final Log logger = LogFactory
            .getLog(InputProductController.class);

    @Override
    public ModelAndView handleRequest(HttpServletRequest request,
            HttpServletResponse response) throws Exception {
        logger.info("InputProductController called");
        return new ModelAndView("/WEB-INF/jsp/ProductForm.jsp");
    }
}
```

InputProductController 类的 handleRequest 方法只是返回一个 ModelAndView，包含一个视图，且没有模型。因此，该请求将被转发到/WEB-INF/jsp/ProductForm.jsp 页面。

清单 3.4　SaveProductController 类

```
package controller;
import javax.servlet.http.HttpServletRequest;
import javax.servlet.http.HttpServletResponse;
import org.apache.commons.logging.Log;
import org.apache.commons.logging.LogFactory;
import org.springframework.web.servlet.ModelAndView;
import org.springframework.web.servlet.mvc.Controller;
import domain.Product;
```

```
import form.ProductForm;

public class SaveProductController implements Controller {

    private static final Log logger = LogFactory
            .getLog(SaveProductController.class);

    @Override
    public ModelAndView handleRequest(HttpServletRequest request,
            HttpServletResponse response) throws Exception {
        logger.info("SaveProductController called");
        ProductForm productForm = new ProductForm();
        // populate action properties
        productForm.setName(request.getParameter("name"));
        productForm.setDescription(request.getParameter(
                "description"));
        productForm.setPrice(request.getParameter("price"));

        // create model
        Product product = new Product();
        product.setName(productForm.getName());
        product.setDescription(productForm.getDescription());
        try {
            product.setPrice(
                    Float.parseFloat(productForm.getPrice()));
        } catch (NumberFormatException e) {
        }

        // insert code to save Product

        return new ModelAndView("/WEB-INF/jsp/ProductDetails.jsp",
                "product", product);
    }
}
```

SaveProductController 类的 handleRequest 方法中，首先用请求参数创建一个 ProductForm 对象；然后，它根据 ProductForm 对象创建 Product 对象。由于 ProductForm 的 price 属性是一个字符串，而其在 Product 类对应的是一个 float，此处类型转换是必要的。在第 4 章中，我们将学习在 Spring MVC 中如何省去 ProductForm 对象，使事情变得更省力。

SaveProductController 的 handleRequest 方法最后返回的 ModelAndView 模型包含了视图的路径、模型名称以及模型（product 对象）。该模型将提供给目标视图，用于界面显示。

3.4.4 View 类

springmvc-intro1 应用程序中包含两个 JSP 页面：ProductForm.jsp 页面（代码见清单 3.5）和 ProductDetails.jsp 页面（见清单 3.6）。

清单 3.5　ProductForm.jsp 页面

```
<!DOCTYPE HTML>
<html>
<head>
<title>Add Product Form</title>
<style type="text/css">@import url(css/main.css);</style>
</head>
<body>

<div id="global">
<form action="save- Product" method="post">
    <fieldset>
        <legend>Add a product</legend>
        <label for="name">Product Name: </label>
        <input type="text" id="name" name="name" value=""
            tabindex="1">
        <label for="description">Description: </label>
        <input type="text" id="description" name="description"
            tabindex="2">
        <label for="price">Price: </label>
        <input type="text" id="price" name="price" tabindex="3">
        <div id="buttons">
            <label for="dummy"> </label>
            <input id="reset" type="reset" tabindex="4">
            <input id="submit" type="submit" tabindex="5"
                value="Add Product">
        </div>
    </fieldset>
</form>
</div>
</body>
</html>
```

此处不适合讨论 HTML 和 CSS，但需要强调的是清单 3.5 中的 HTML 是经过适当设计的，并且没有使用<table>来布局输入字段。

清单 3.6　ProductDetails.jsp 页面

```
<!DOCTYPE HTML>
<html>
```

```
<head>
<title>Save Product</title>
<style type="text/css">@import url(css/main.css);</style>
</head>
<body>
<div id="global">
    <h4>The product has been saved.</h4>
    <p>
        <h5>Details:</h5>
        Product Name: ${product.name}<br/>
        Description: ${product.description}<br/>
        Price: $${product.price}
    </p>
</div>
</body>
</html>
```

ProductDetails.jsp 页面通过模型属性名"product"来访问由 SaveProductController 传入的 Product 对象。这里用 JSP 表达式语言来显示 Product 对象的各种属性。我们将在第 8 章中学习 JSP EL。

3.4.5 测试应用

现在，在浏览器中输入如下 URL 来测试应用：

```
http://localhost:8080/springmvc-introl/input-product
```

会看到类似图 3.2 的产品表单页面，在空字段中输入相应的值后单击"Add Product（添加产品）"按钮，会在下一页中看到产品属性。

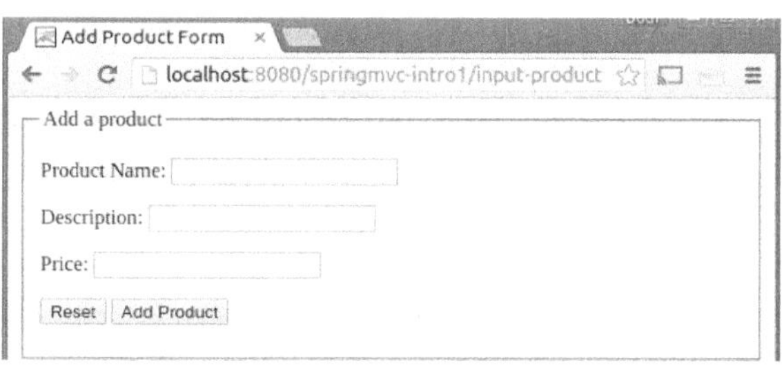

图 3.2 springmvc-intro1 的产品表单

3.5 视图解析器

Spring MVC 中的视图解析器负责解析视图。可以通过在配置文件中定义一个 ViewResolver（如下）来配置视图解析器。

```
<bean id="viewResolver" class="org.springframework.web.servlet.
➥view.InternalResourceViewResolver">
    <property name="prefix" value="/WEB-INF/jsp/"/>
    <property name="suffix" value=".jsp"/>
</bean>
```

视图解析器配置有前缀和后缀两个属性。这样一来，view 路径将缩短。例如，仅需提供“myPage”，而不必再将视图路径设置为/WEB-INF/jsp/myPage.jsp，视图解析器将会自动增加前缀和后缀。

以 springmvc-intro2 应用为例，该例子同 springmvc-intro1 应用类似，但调整了配置文件的名称和路径。此外，它还配置了默认的视图解析器，为所有视图路径添加前缀和后缀，如图 3.3 所示。

springmvc-intro2 中，Spring MVC 的配置文件被重命名为 springmvc-config.xml 中，并移动到/ WEB-INF/config 目录下。为了让 Spring MVC 可以正确加载到该配置文件，需要将文件路径配置到 Spring MVC 的 Dispatcher servlet。清单 3.7 显示了 springmvc-intro2 应用的部署描述符（web.xml 文件）。

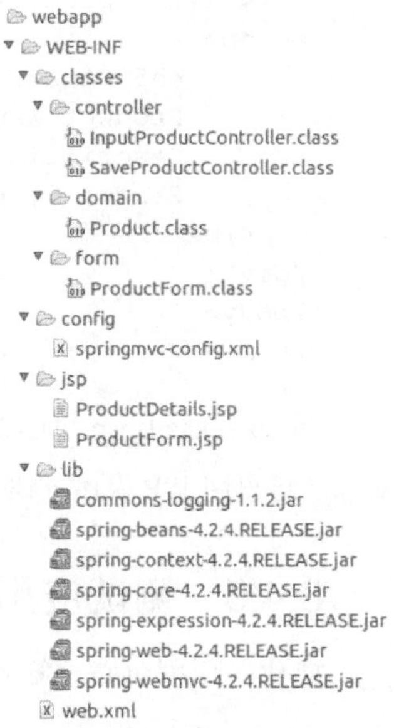

```
webapp
▾ WEB-INF
  ▾ classes
    ▾ controller
        InputProductController.class
        SaveProductController.class
    ▾ domain
        Product.class
    ▾ form
        ProductForm.class
  ▾ config
      springmvc-config.xml
  ▾ jsp
      ProductDetails.jsp
      ProductForm.jsp
  ▾ lib
      commons-logging-1.1.2.jar
      spring-beans-4.2.4.RELEASE.jar
      spring-context-4.2.4.RELEASE.jar
      spring-core-4.2.4.RELEASE.jar
      spring-expression-4.2.4.RELEASE.jar
      spring-web-4.2.4.RELEASE.jar
      spring-webmvc-4.2.4.RELEASE.jar
    web.xml
```

图 3.3　springmvc-intro2 文件结构

清单 3.7　springmvc-intro2 应用的部署描述符

```
<?xml version="1.0" encoding="UTF-8"?>
<web-app version="3.1"
    xmlns="http://xmlns.jcp.org/xml/ns/javaee"
    xmlns:xsi="http://www.w3.org/2001/XMLSchema-instance"
    xsi:schemaLocation="http://xmlns.jcp.org/xml/ns/javaee
➥http://xmlns.jcp.org/xml/ns/javaee/web-app_3_1.xsd">

    <servlet>
        <servlet-name>springmvc</servlet-name>
        <servlet-class>
            org.springframework.web.servlet.DispatcherServlet
        </servlet-class>
        <init-param>
            <param-name>contextConfigLocation</param-name>
            <param-value>
                /WEB-INF/config/springmvc-config.xml
            </param-value>
        </init-param>
        <load-on-startup>1</load-on-startup>
```

```
    </servlet>

    <servlet-mapping>
        <servlet-name>springmvc</servlet-name>
        <url-pattern>/</url-pattern>
    </servlet-mapping>
</web-app>
```

需要特别注意的是 web.xml 文件中的 init-param 元素。要使用非默认配置文件的命名和路径，需要使用名为 contextConfigLocation 的 init-param，其值应为配置文件在应用中的相对路径（见清单 3.8）。

清单 3.8　springmvc-intro2 的配置文件

```
<?xml version="1.0" encoding="UTF-8"?>
<beans xmlns="http://www.springframework.org/schema/beans"
  xmlns:xsi="http://www.w3.org/2001/XMLSchema-instance"
  xsi:schemaLocation="http://www.springframework.org/schema/beans
  http://www.springframework.org/schema/beans/spring-beans.xsd">

    <bean name="/input-product"
            class="controller.InputProductController"/>
    <bean name="/save-product"
            class="controller.SaveProductController"/>
    <bean id="viewResolver"
            class="org.springframework.web.servlet.view.
➥InternalResourceViewResolver">
        <property name="prefix" value="/WEB-INF/jsp/"/>
        <property name="suffix" value=".jsp"/>
    </bean>
</beans>
```

在浏览器中输入如下 URL，测试 app03b 应用：

```
http://localhost:8080/springmvc-intro2/input-product
```

即可看到如图 3.2 所示的表单页面。

3.6　小结

本章是 Spring MVC 的入门介绍。我们学习了如何开发一个类似第 2 章的简单应用。在 Spring MVC 中，我们无需编写自己的 Dispatcher servlet，并通过实现控制器接口来编写控制器。这是传统风格的控制器。从 Spring 2.5 版本开始，Spring 提供了一个更好的开发控制器的方式，如采用注解。第 4 章会深入介绍这种风格的控制器。

第 4 章 基于注解的控制器

在第 3 章中，我们创建了两个采用传统风格控制器的 Spring MVC 应用程序。其控制器是实现了 Controller 接口的类。Spring 2.5 版本引入了一个新途径：使用控制器注释类型。本章介绍了基于注解的控制器，以及各种对应用程序有用的注释类型。

4.1 Spring MVC 注解类型

使用基于注解的控制器的几个优点。其一，一个控制器类可以处理多个动作（而实现了 Controller 接口的一个控制器只能处理一个动作）。这就允许将相关的操作写在同一个控制器类中，从而减少应用程序中类的数量。

其二，基于注解的控制器的请求映射不需要存储在配置文件中。使用 RequestMapping 注释类型，可以对一个方法进行请求处理。

Controller 和 RequestMapping 注释类型是 Spring MVC API 最重要的两个注解类型。本章重点介绍这两个，并简要介绍了一些其他不太流行的注解类型。

4.1.1 Controller 注解类型

org.springframework.stereotype.Controller 注解类型用于指示 Spring 类的实例是一个控制器。下面是一个带注解@Controller 的例子。

```
package com.example.controller;

import org.springframework.stereotype;
...

@Controller
public class CustomerController {
```

```
    // request-handling methods here
}
```

Spring 使用扫描机制来找到应用程序中所有基于注解的控制器类。为了保证 Spring 能找到你的控制器，需要完成两件事情。首先，需要在 Spring MVC 的配置文件中声明 spring-context，如下所示：

```
<beans
    ...
    xmlns:context="http://www.springframework.org/schema/context"
    ...
>
```

然后，需要应用<component-scan/>元素，如下所示：

```
<context:component-scan base-package="basePackage"/>
```

请在<component-scan/>元素中指定控制器类的基本包。例如，若所有的控制器类都在 com.example.controller 及其子包下，则需要写一个如下所示的<component-scan/>元素：

```
<context:component-scan base-package="com.example.controller"/>
```

现在，整个配置文件看上去如下所示：

```
<?xml version="1.0" encoding="UTF-8"?>
<beans xmlns="http://www.springframework.org/schema/beans"
    xmlns:xsi="http://www.w3.org/2001/XMLSchema-instance"
    xmlns:p="http://www.springframework.org/schema/p"
    xmlns:context="http://www.springframework.org/schema/context"
    xsi:schemaLocation="
        http://www.springframework.org/schema/beans
        http://www.springframework.org/schema/beans/spring-beans.xsd
        http://www.springframework.org/schema/context
        http://www.springframework.org/schema/context/spring-
       context.xsd">

    <context:component-scan base-package="com.example.controller"/>

    <!-- ... -->
</beans>
```

请确保所有控制器类都在基本包下，并且不要指定一个太广泛的基本包（如指定 com.example，而非 com.example.controller，前者就更广泛），因为这会使得 Spring MVC 扫描了无关的包。

4.1.2 RequestMapping 注解类型

现在，我们需要在控制类的内部为每一个动作开发相应的处理方法。要让 Spring 知道用哪一种方法来处理它的动作，需要使用 org.springframework.web.bind.annotation.RequestMapping 注解类型映射的 URI 与方法。

RequestMapping 注解类型的作用同其名字所暗示的：映射一个请求和一种方法。可以使用@RequestMapping 注解一种方法或类。

一个采用@RequestMapping 注解的方法将成为一个请求处理方法，并由调度程序在接收到对应 URL 请求时调用。

下面是一个 RequestMapping 注解方法的控制器类。

```
package com.example.controller;
import org.springframework.stereotype.Controller;
import org.springframework.web.bind.annotation.RequestMapping;
...

@Controller
public class CustomerController {

    @RequestMapping(value = "/input-customer ")
    public String inputCustomer() {

        // do something here

        return "CustomerForm";
    }
}
```

使用 RequestMapping 注解的 value 属性将 URI 映射到方法。在上面的例子中，我们将 input-customer 映射到 inputCustomer 方法。这样，可以使用如下 URL 访问 inputCustomer 方法。

```
http://domain/context/input-customer
```

由于 value 属性是 RequestMapping 注释的默认属性，因此，若只有唯一的属性，则可以省略属性名称。换句话说，如下两个标注含义相同。

```
@RequestMapping(value = "/input-customer ")
@RequestMapping("/input-customer ")
```

但如果有多个属性时，就必须写入 value 属性名称。

请求映射的值可以是一个空字符串，此时该方法被映射到以下网址：

```
http://domain/context
```

RequestMapping 除了具有 value 属性外，还有其他属性。例如，method 属性用来指示该方法仅处理哪些 HTTP 方法。

例如，仅当在 HTTP POST 或 PUT 方法时，才访问到下面的 ProcessOrder 方法。

```
...
import org.springframework.stereotype.Controller;
import org.springframework.web.bind.annotation.RequestMapping;
import org.springframework.web.bind.annotation.RequestMethod;
...
    @RequestMapping(value="/process-order",
            method={RequestMethod.POST, RequestMethod.PUT})
    public String processOrder() {

        // do something here

        return "OrderForm";
    }
```

若 method 属性只有一个 HTTP 方法值，则无需花括号。例如，

```
@RequestMapping(value="/process-order", method=RequestMethod.POST)
```

如果没有指定 method 属性值，则请求处理方法可以处理任意 HTTP 方法。

此外，RequestMapping 注解类型也可以用来注解一个控制器类，如下所示：

```
import org.springframework.stereotype.Controller;
...

@Controller
@RequestMapping(value="/customer")
public class CustomerController {
```

在这种情况下，所有的方法都将映射为相对于类级别的请求。例如，下面的 deleteCustomer 方法。

```
...
import org.springframework.stereotype.Controller;
import org.springframework.web.bind.annotation.RequestMapping;
import org.springframework.web.bind.annotation.RequestMethod;
...
@Controller
@RequestMapping("/customer")
public class CustomerController {

    @RequestMapping(value="/delete",
            method={RequestMethod.POST, RequestMethod.PUT})
    public String deleteCustomer() {

        // do something here
```

```
        return ...;
}
```

由于控制器类的映射使用“/customer”，而 deleteCustomer 方法映射为“/delete”，则如下 URL 会映射到该方法上。

```
http://domain/context/customer/delete
```

4.2　编写请求处理方法

每个请求处理方法可以有多个不同类型的参数，以及一个多种类型的返回结果。例如，如果在请求处理方法中需要访问 HttpSession 对象，则可以添加的 HttpSession 作为参数，Spring 会将对象正确地传递给方法。

```
@RequestMapping("/uri")
public String myMethod(HttpSession session) {
    ...
    session.addAttribute(key, value);
    ...
}
```

或者，如果需要访问客户端语言环境和 HttpServletRequest 对象，则可以在方法签名上包括这样的参数：

```
@RequestMapping("/uri")
public String myOtherMethod(HttpServletRequest request,
        Locale locale) {
    ...
    // access Locale and HttpServletRequest here
    ...
}
```

下面是可以在请求处理方法中出现的参数类型：

- javax.servlet.ServletRequest 或 javax.servlet.http.HttpServletRequest。
- javax.servlet.ServletResponse 或 javax.servlet.http.HttpServletResponse。
- javax.servlet.http.HttpSession。
- org.springframework.web.context.request.WebRequest 或 org.springframework.web.context.request.NativeWebRequest。

- java.util.Locale。
- java.io.InputStream 或 java.io.Reader。
- java.io.OutputStream 或 java.io.Writer。
- java.security.Principal。
- HttpEntity<?>paramters
- java.util.Map/org.springframework.ui.Model /。
- org.springframework.ui.ModelMap。
- org.springframework.web.servlet.mvc.support.RedirectAttributes。
- org.springframework.validation.Errors /。
- org.springframework.validation.BindingResult。
- 命令或表单对象。
- org.springframework.web.bind.support.SessionStatus。
- org.springframework.web.util.UriComponentsBuilder。
- 带@PathVariable、@MatrixVariable、@RequestParam、@RequestHeader、@RequestBody 或@RequestPart 注释的对象。

特别重要的是 org.springframework.ui.Model 类型。这不是一个 Servlet API 类型，而是一个包含 Map 的 Spring MVC 类型。每次调用请求处理方法时，Spring MVC 都创建 Model 对象并将其 Map 注入到各种对象。

请求处理方法可以返回如下类型的对象：

- ModelAndView。
- Model。
- 包含模型的属性的 Map。
- View。
- 代表逻辑视图名的 String。
- void。

- 提供对 Servlet 的访问，以响应 HTTP 头部和内容 HttpEntity 或 ResponseEntity 对象。
- Callable。
- DeferredResult。
- 其他任意类型，Spring 将其视作输出给 View 的对象模型。

本章后续会展示一个例子，以进一步学习如何开发一个请求处理方法。

4.3 应用基于注解的控制器

本章的示例应用 annotated1 基于第 2 章和第 3 章的例子重写，展示了包含有两个请求处理方法的一个控制器类。

annotated1 和前面的应用程序间的主要区别在于，annotated1 的控制器类增加了注解 @Controller。此外，Spring 配置文件也增加了一些元素，后续小节中会详细介绍。

4.3.1 目录结构

图 4.1 展示了 annotated1 的目录结构。注意，annotated1 中只有一个控制器类，而不是两个，同时新增了一个名为 index.html 的 HTML 文件，以便 Spring MVC Servlet 的 URL 模式设置为“/”时，依然可以访问静态资源。

4.3.2 配置文件

annotated1 有两个配置文件。第一个为部署描述符（web.xml 文件）中注册 Spring MVC 的 Dispatcher Servlet。第二个为 springmvc-config.xml，即 Spring MVC 的配置文件。

清单 4.1 和清单 4.2 分别展示部署描述符和 Spring MVC 的配置文件。

清单 4.1 annotated1(web.xml)的部署描述符

```
<?xml version="1.0" encoding="UTF-8"?>
<web-app version="3.1"
    xmlns="http://xmlns.jcp.org/xml/ns/javaee"
    xmlns:xsi="http://www.w3.org/2001/XMLSchema-instance"
    xsi:schemaLocation="http://xmlns.jcp.org/xml/ns/javaee
        http://xmlns.jcp.org/xml/ns/javaee/web-app_3_1.xsd">

    <servlet>
```

```
        <servlet-name>springmvc</servlet-name>
        <servlet-class>
            org.springframework.web.servlet.DispatcherServlet
        </servlet-class>
        <init-param>
            <param-name>contextConfigLocation</param-name>
            <param-value>
                /WEB-INF/config/springmvc-config.xml
            </param-value>
        </init-param>
        <load-on-startup>1</load-on-startup>
    </servlet>

    <servlet-mapping>
        <servlet-name>springmvc</servlet-name>
        <url-pattern>/</url-pattern>
    </servlet-mapping>
</web-app>
```

图 4.1 为 annotated1 的目录结构。

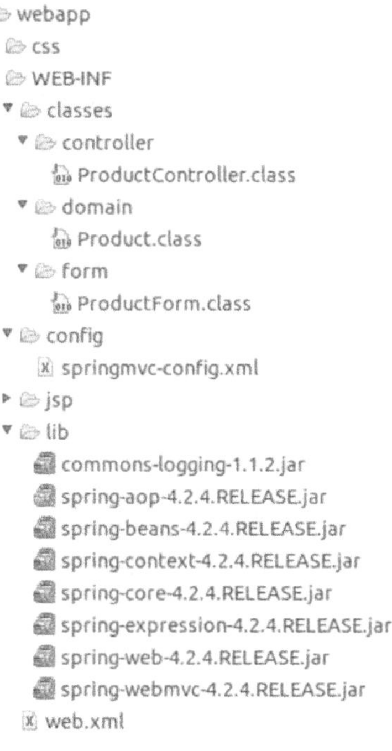

图 4.1 annotated1 的目录结构

另外，在部署描述符中的<servlet-mapping/>元素，Spring MVC 的 dispatcher-servlet 的 URL

模式设置为“/”，当 URL 模式设置为“/”时，意味着所有请求（包括那些用于静态资源）都被映射到 Dispatcher Servlet。为了正确处理静态资源，需要在 Spring MVC 配置文件中添加一些<resources/>元素。

清单 4.2　springmvc-config.xml 文件

```
<?xml version="1.0" encoding="UTF-8"?>
<beans xmlns="http://www.springframework.org/schema/beans"
    xmlns:xsi="http://www.w3.org/2001/XMLSchema-instance"
    xmlns:p="http://www.springframework.org/schema/p"
    xmlns:mvc="http://www.springframework.org/schema/mvc"
    xmlns:context="http://www.springframework.org/schema/context"
    xsi:schemaLocation="
        http://www.springframework.org/schema/beans
        http://www.springframework.org/schema/beans/spring-beans.xsd
        http://www.springframework.org/schema/mvc
        http://www.springframework.org/schema/mvc/spring-mvc.xsd
        http://www.springframework.org/schema/context
        http://www.springframework.org/schema/context/spring-context.xsd">
    <context:component-scan base-package="controller"/>
    <mvc:annotation-driven/>
    <mvc:resources mapping="/css/**" location="/css/"/>
    <mvc:resources mapping="/*.html" location="/"/>

    <bean id="viewResolver"
        class="org.springframework.web.servlet.view.
➥InternalResourceViewResolver">
        <property name="prefix" value="/WEB-INF/jsp/"/>
        <property name="suffix" value=".jsp"/>
    </bean>
</beans>
```

清单 4.2（Spring MVC 的配置文件）中最主要的是<component-scan/>元素。这是要指示 Spring MVC 扫描目标包中的类，本例是 controller 包。接下去是一个<annotation-driven/>元素和两个<resources/>元素。<annotation-driven/>元素做了很多的事情，其中包括注册用于控制器注解的 bean 对象。<resources/>元素则指示 Spring MVC 哪些静态资源需要单独处理（不通过 Dispatcher Servlet）。

在清单 4.2 的配置文件中，有两个<resources/>元素。第一个确保在/ CSS 目录下的所有文件可见，第二个允许显示所有的.html 文件。

注意

如果没有<annotation-driven/>，<resources/>元素会阻止任意控制器被调用。若不需要使用 resources，则不需要<annotation-driven/>元素。

4.3.3　Controller 类

如前所述，使用 Controller 注释类型的一个优点在于：一个控制器类可以包含多个请求处理方法。如清单 4.3 所示，ProductController 类中有 inputProduct 和 saveProduct 两个方法。

清单 4.3　ProductController 类

```
package controller;
import java.match.Bigoecimal
import org.apache.commons.logging.Log;
import org.apache.commons.logging.LogFactory;
import org.springframework.stereotype.Controller;
import org.springframework.ui.Model;
import org.springframework.web.bind.annotation.RequestMapping;
import domain.Product;
import form.ProductForm;

@Controller
public class ProductController {

    private static final Log logger =
       LogFactory.getLog(ProductController.class);

    @RequestMapping(value="/input-product")
    public String inputProduct() {
        logger.info("inputProduct called");
        return "ProductForm";
    }

    @RequestMapping(value="/save-product")
    public String saveProduct(ProductForm productForm, Model model){
        logger.info("saveProduct called");
        // no need to create and instantiate a ProductForm
        // create Product
        Product product = new Product();
        product.setName(productForm.getName());
        product.setDescription(productForm.getDescription());
        try {
            product.setPrice(new BigDecimal(productForm.getPrice()));
        } catch (NumberFormatException e) {
```

```
            }

            // add product
            model.addAttribute("product", product);
            return "ProductDetails";
    }
}
```

其中，ProductController 的 saveProduct 方法的第二个参数是 org.springframework.ui.Model 类型。无论是否会使用，Spring MVC 都会在每一个请求处理方法被调用时创建一个 Model 实例，用于增加需要显示在视图中的属性。例如，通过调用 model.addAttribute 来添加 Product 实例：

```
model.addAttribute("product", product);
```

Product 实例就可以像被添加到 HttpServletRequest 中那样访问了。

4.3.4 View

annotated1 也有与前面章节示例类似的两个视图：ProductForm.jsp 页面（见清单 4.4）和 ProductDetails.jsp 页面（见清单 4.5）。

清单 4.4　ProductForm.jsp 页面

```
<!DOCTYPE HTML>
<html>
<head>
<title>Add Product Form</title>
<style type="text/css">@import url(css/main.css);</style>
</head>
<body>

<div id="global">
<form action="save-product" method="post">
    <fieldset>
        <legend>Add a product</legend>
        <p>
            <label for="name">Product Name: </label>
            <input type="text" id="name" name="name"
                 tabindex="1">
        </p>
        <p>
            <label for="description">Description: </label>
            <input type="text" id="description"
                name="description" tabindex="2">
        </p>
```

```
        <p>
            <label for="price">Price: </label>
            <input type="text" id="price" name="price"
                tabindex="3">
        </p>
        <p id="buttons">
            <input id="reset" type="reset" tabindex="4">
            <input id="submit" type="submit" tabindex="5"
                value="Add Product">
        </p>
    </fieldset>
</form>
</div>
</body>
</html>
```

清单 4.5 ProductDetails.jsp 页面

```
<!DOCTYPE HTML>
<html>
<head>
<title>Save Product</title>
<style type="text/css">@import url(css/main.css);</style>
</head>
<body>
<div id="global">
    <h4>The product has been saved.</h4>
    <p>
        <h5>Details:</h5>
        Product Name: ${product.name}<br/>
        Description: ${product.description}<br/>
        Price: $${product.price}
    </p>
</div>
</body>
</html>
```

4.3.5 测试应用

在浏览器中输入如下 URL 来测试 annotated1。

```
http://localhost:8080/annotated1/input-product
```

浏览器会显示 Product 表单，如图 4.2 所示。

单击“Add Product”按钮，会调用 saveProduct 方法。

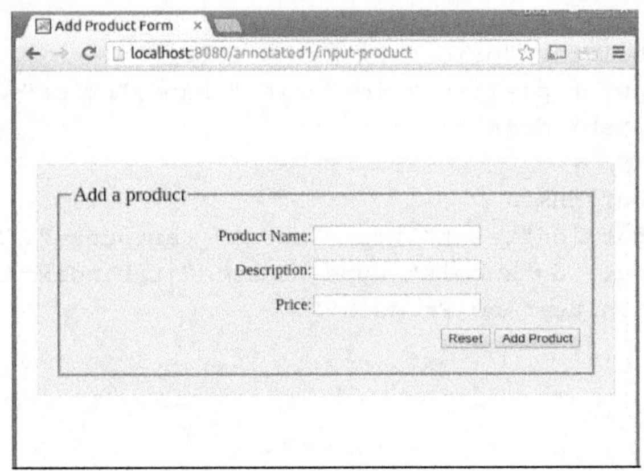

图 4.2　Product 表单

4.4　应用@Autowired 和@Service 进行依赖注入

使用 Spring 框架的一个好处是容易进行依赖注入。毕竟，Spring 框架一开始就是一个依赖注入容器。将依赖注入到 Spring MVC 控制器的最简单方法是，通过注解@Autowired 到字段或方法。Autowired 注解类型属于 org.springframework.beans.factory.annotation 包。

此外，为了能作为依赖注入，类必须要注明为@Service。该类型是 org.springframework.stereotype 包的成员。Service 注解类型指示类是一个服务。此外，在配置文件中，还需要添加一个<component-scan/>元素来扫描依赖基本包。

```
<context:component-scan base-package="dependencyPackage"/>
```

下面以 annotated2 应用进一步说明 Spring MVC 如何应用依赖注入。在 annotated2 应用程序中，ProductController 类（见清单 4.6）已经不同于 annotated1 中的类。

清单 4.6　annotated2 中的 ProductController 类

```
package controller;
import java.math.BigDecimal;
import org.apache.commons.logging.Log;
import org.apache.commons.logging.LogFactory;
import org.springframework.beans.factory.annotation.Autowired;
import org.springframework.stereotype.Controller;
import org.springframework.ui.Model;
import org.springframework.web.bind.annotation.PathVariable;
import org.springframework.web.bind.annotation.RequestMapping;
```

```
import org.springframework.web.bind.annotation.RequestMethod;
import org.springframework.web.servlet.mvc.support.RedirectAttributes;
import domain.Product;
import form.ProductForm;
import service.ProductService;

@Controller
public class ProductController {
    private static final Log logger = LogFactory
            .getLog(ProductController.class);

    @Autowired
    private ProductService productService;

    @RequestMapping(value = "/input-product ")
    public String inputProduct() {
        logger.info("inputProduct called");
        return "ProductForm";
    }

    @RequestMapping(value = "/save-product ", method = RequestMethod.POST)
    public String saveProduct(ProductForm productForm,
            RedirectAttributes redirectAttributes) {
        logger.info("saveProduct called");
        // no need to create and instantiate a ProductForm
        // create Product
        Product product = new Product();
        product.setName(productForm.getName());
        product.setDescription(productForm.getDescription());
        try {
            product.setPrice(new BigDecimal(productForm.getPrice()));
        } catch (NumberFormatException e) {
        }

        // add product
        Product savedProduct = productService.add(product);

        redirectAttributes.addFlashAttribute("message",
                "The product was successfully added.");
        return "redirect:/view-product/" + savedProduct.getId();
    }

    @RequestMapping(value = "/view-product/{id}")
    public String viewProduct(@PathVariable Long id, Model model) {
        Product product = productService.get(id);
```

```
        model.addAttribute("product", product);
        return "ProductPetails";
    }
}
```

与 annotated1 中相比，annotated2 中的 ProductController 类做了一系列的调整。首先是在如下的私有字段上增加了@Autowired 注解：

```
@Autowired
private ProductService productService
```

ProductService 是一个提供各种处理产品的方法的接口。为 productService 字段添加@Autowired 注解会使 ProductService 的一个实例被注入到 ProductController 实例中。

清单 4.7 和清单 4.8 分别显示了 ProductService 接口及其实现类 ProductServiceImpl。注意，为了使类能被 Spring 扫描到，必须为其标注@Service。

清单 4.7　ProductService 接口

```
package service
import domain.Product;
public interface ProductService {
    Product add(Product product);
    Product get(long id);
}
```

清单 4.8　ProductServiceImpl 类

```
package service;
import java.math.BigDecimal;
import java.util.HashMap;
import java.util.Map;
import java.util.concurrent.atomic.AtomicLong;
import org.springframework.stereotype.Service;
import domain.Product;

@Service
public class ProductServiceImpl implements ProductService {

    private Map<Long, Product> products =
            new HashMap<Long, Product>();
    private AtomicLong generator = new AtomicLong();

    public ProductServiceImpl() {
        Product product = new Product();
```

```
        product.setName("JX1 Power Drill");
        product.setDescription(
                "Powerful hand drill, made to perfection");
        product.setPrice(new BigDecimal(129.99));

        add(product);
    }

    @Override
    public Product add(Product product) {
        long newId = generator.incrementAndGet();
        product.setId(newId);
        products.put(newId, product);
        return product;
    }

    @Override
    public Product get(long id) {
        return products.get(id);
    }
}
```

如清单4.9所示，annotated2的Spring MVC配置文件中有两个<component-scan/>元素：一个用于扫描控制器类，另一个用于扫描服务类。

清单4.9 Spring MVC配置文件

```
<?xml version="1.0" encoding="UTF-8"?>
<beans xmlns="http://www.springframework.org/schema/beans"
    xmlns:xsi="http://www.w3.org/2001/XMLSchema-instance"
    xmlns:p="http://www.springframework.org/schema/p"
    xmlns:mvc="http://www.springframework.org/schema/mvc"
    xmlns:context="http://www.springframework.org/schema/context"
    xsi:schemaLocation="
        http://www.springframework.org/schema/beans
        http://www.springframework.org/schema/beans/spring-beans.xsd
        http://www.springframework.org/schema/mvc
        http://www.springframework.org/schema/mvc/spring-mvc.xsd
        http://www.springframework.org/schema/context
        http://www.springframework.org/schema/context/springcontext.xsd">

    <context:component-scan base-package="controller"/>
    <context:component-scan base-package="service"/>
    <mvc:annotation-driven/>
    <mvc:resources mapping="/css/**" location="/css/"/>
    <mvc:resources mapping="/*.html" location="/"/>

    <bean id="viewResolver"
```

```
        class="org.springframework.web.servlet.view.InternalResourceViewResolver">
            <property name="prefix" value="/WEB-INF/jsp/"/>
            <property name="suffix" value=".jsp"/>
    </bean>
</beans>
```

4.5　重定向和 Flash 属性

作为一名经验丰富的 servlet/ JSP 程序员，必须知道转发和重定向的区别。转发比重定向快，因为重定向经过客户端，而转发没有。但是，有时采用重定向更好，若需要重定向到一个外部网站，则无法使用转发。

使用重定向的另一个场景是避免在用户重新加载页面时再次调用同样的动作。例如，在 annotated1 中，当提交产品表单时，saveProduct 方法将被调用，并执行相应的动作。在一个真实的应用程序中，这可能包括将所述产品加入到数据库中。但是，如果在提交表单后重新加载页面，saveProduct 就会被再次调用，同样的产品将可能被再次添加。为了避免这种情况，提交表单后，你可能更愿意将用户重定向到一个不同的页面。这个网页任意重新加载都没有副作用。例如，在 annotated1 中，可以在提交表单后，将用户重定向到一个 ViewProduct 页面。

在 annotated2 中，ProductController 类中的 saveProduct 方法以如下所示的行结束：

```
return "redirect:/view-product/" + savedProduct.getId();
```

这里，使用重定向而不是转发来防止当用户重新加载页面时 saveProduct 被二次调用。

使用重定向的一个不便的地方是：无法轻松地传值给目标页面。而采用转发，则可以简单地将属性添加到 Model，使得目标视图可以轻松访问。由于重定向经过客户端，所以 Model 中的一切都在重定向时丢失。好在，Spring 3.1 版本以及更高版本通过 Flash 属性提供了一种供重定向传值的方法。

要使用 Flash 属性，必须在 Spring MVC 配置文件中有一个<annotation-driven/>元素。然后，还必须在方法上添加一个新的参数类型 org.springframework.web.servlet.mvc.support.RedirectAttributes。清单 4.10 展示了更新后的 saveProduct 方法。

清单 4.10　使用 Flash 属性

```
@RequestMapping(value = "save-product", method = RequestMethod.POST)
public String saveProduct(ProductForm productForm,
        RedirectAttributes redirectAttributes) {
    logger.info("saveProduct called");
```

```
        // no need to create and instantiate a ProductForm
        // create Product
        Product product = new Product();
        product.setName(productForm.getName());
        product.setDescription(productForm.getDescription());
        try {
            product.setPrice(new BigDecimal(productForm.getPrice()));
        } catch (NumberFormatException e) {
        }

        // add product
        Product savedProduct = productService.add(product);

        redirectAttributes.addFlashAttribute("message",
                "The product was successfully added.");

        return "redirect:/ view_product /" + savedProduct.getId();
    }
```

4.6　请求参数和路径变量

请求参数和路径变量都可以用于发送值给服务器。二者都是 URL 的一部分。请求参数采用 key=value 形式，并用“&”分隔。例如，下面的 URL 带有一个名为 productId 的请求参数，其值为 3：

```
http://localhost:8080/annotated2/view-product?productId=3
```

在传统的 servlet 编程中，可以使用 HttpServletRequest 的 getParameter 方法来获取一个请求参数值：

```
String productId = httpServletRequest.getParameter("productId");
```

Spring MVC 提供了一个更简单的方法来获取请求参数值：使用 org.springframework.web.bind.annotation.RequestParam 注解类型来注解方法参数。例如，下面的方法包含了一个获取请求参数 productId 值的参数。

```
public void sendProduct(@RequestParam int productId)
```

正如你所看到的，@RequestParam 注解的参数类型不一定是字符串。

路径变量类似请求参数，但没有 key 部分，只是一个值。例如，在 annotated2 中，view-product 动作映射到如下 URL：

```
/view-product/productId
```

其中的 productId 是表示产品标识符的整数。在 Spring MVC 中，productId 称为路径变量，用来发送一个值到服务器。

清单 4.11 中的 viewProduct 方法演示了一个路径变量的使用。

清单 4.11　使用路径变量

```
@RequestMapping(value = "/view-product/{id}")
public String viewProduct(@PathVariable Long id, Model model) {
    Product product = productService.get(id);
    model.addAttribute("product", product);
    return "ProductView";
}
```

为了使用路径变量，首先需要在 RequestMapping 注解的值属性中添加一个变量，该变量必须放在花括号之间。例如，下面的 RequestMapping 注解定义了一个名为 id 的路径变量：

```
@RequestMapping(value = "/view-product/{id}")
```

然后，在方法签名中添加一个同名变量，并加上@PathVariable 注解。注意清单 4.11 中 viewProduct 的方法签名。当该方法被调用时，请求 URL 的 id 值将被复制到路径变量中，并可以在方法中使用。路径变量的类型可以不是字符串。Spring MVC 将尽力转换成一个非字符串类型。这个 Spring MVC 的强大功能会在第 5 章中详细讨论。

可以在请求映射中使用多个路径变量。例如，下面定义了 userId 和 orderId 两个路径变量。

```
@RequestMapping(value = "/view-product/{userId}/{orderId}")
```

请直接在浏览器中输入如下 URL，来测试 viewProduct 方法的路径变量。

```
http://localhost:8080/annotated2/view-product/1
```

有时，使用路径变量时会遇到一个小问题：在某些情况下，浏览器可能会误解路径变量。考虑下面的 URL。

```
http://example.com/context/abc
```

浏览器会（正确）认为 abc 是一个动作。任何静态文件路径的解析，如 CSS 文件，将使用 http://example.com/context 作为基本路径。这就是说，若服务器发送的网页包含如下 img 元素：

```
<img src="logo.png"/>
```

该浏览器将试图通过 http://example.com/context/logo.png 来加载 logo.png。

然而，若一个应用程序被部署为默认上下文（默认上下文路径是一个空字符串），则对于同一个目标的 URL，会是这样的：

```
http://example.com/abc
```

下面是带有路径变量的 URL：

```
http://example.com/abc/1
```

在这种情况下，浏览器会认为 abc 是上下文，没有动作。如果在页面中使用<img src="logo.png"/>，浏览器将试图通过 http://example.com/abc/logo.png 来寻找图像，并且它将找不到该图像。

好在，我们有一个简单的解决方案，即通过使用 JSTL 标记的 URL（第 8 章中将会详细讨论 JSTL）。标签会通过正确解析 URL 来修复该问题。例如，annotated2 中所有的 JSP 页面导入的所有 CSS，从

```
<style type="text/css">@import url(css/main.css);</style>
```

修改为

```
<style type="text/css">
@import url("<c:url value="/css/main.css"/>");
</style>
```

若程序部署为默认上下文，链接标签会将该 URL 转换成如下所示形式：

```
<style type="text/css">@import url("/css/main.css");</style>
```

若程序不在默认上下文中，则它会被转换成如下形式：

```
<style type="text/css">@import url("/annotated2/css/main.css");
</style>
```

4.7 @ModelAttribute

前面谈到 Spring MVC 在每次调用请求处理方法时，都会创建 Model 类型的一个实例。若打算使用该实例，则可以在方法中添加一个 Model 类型的参数。事实上，还可以使用在方法中添加 ModelAttribute 注解类型来访问 Model 实例。该注解类型也是 org.springframework.web.bind.annotation 包的成员。

可以用@ModelAttribute 来注解方法参数或方法。带@ModelAttribute 注解的方法会将其输入的或创建的参数对象添加到 Model 对象中（若方法中没有显式添加）。例如，Spring MVC

将在每次调用 submitOrder 方法时创建一个 Order 实例。

```
@RequestMapping(method = RequestMethod.POST)
public String submitOrder(@ModelAttribute("newOrder") Order order,
    Model model) {

    ...
}
```

输入或创建的 Order 实例将用 newOrder 键值添加到 Model 对象中。如果未定义键值名，则将使用该对象类型的名称。例如，每次调用如下方法，会使用键值 order 将 Order 实例添加到 Model 对象中。

```
public String submitOrder(@ModelAttribute Order order, Model model)
```

@ModelAttribute 的第二个用途是标注一个非请求的处理方法。被@ModelAttribute 注解的方法会在每次调用该控制器类的请求处理方法时被调用。这意味着，如果一个控制器类有两个请求处理方法，以及一个有@ModelAttribute 注解的方法，该方法的调用次数就会比每个处理请求方法更频繁。

Spring MVC 会在调用请求处理方法之前调用带@ModelAttribute 注解的方法。带@ModelAttribute 注解的方法可以返回一个对象或一个 void 类型。如果返回一个对象，则返回对象会自动添加到 Model 中。

```
@ModelAttribute
public Product addProduct(@RequestParam String productId) {
    return productService.get(productId);
}
```

若方法返回 void，则还必须添加一个 Model 类型的参数，并自行将实例添加到 Model 中，如下面的例子所示。

```
@ModelAttribute
public void populateModel(@RequestParam String id, Model model) er);
    model.addAttribute(new Account(id));
}
```

4.8　小结

在本章中，我们学会了如何编写基于注解的控制器的 Spring MVC 应用，也学会了各种可用来注解类、方法或方法的参数的注解类型。

第 5 章
数据绑定和表单标签库

数据绑定是将用户输入绑定到领域模型的一种特性。有了数据绑定，类型总是为 String 的 HTTP 请求参数，可用于填充不同类型的对象属性。数据绑定使得 form bean（如前面各章中的 ProductForm 实例）变成多余的。

为了高效地使用数据绑定，还需要 Spring 的表单标签库。本章着重介绍数据绑定和表单标签库，并提供范例，展示表单标签库中这些标签的用法。

5.1 数据绑定概览

基于 HTTP 的特性，所有 HTTP 请求参数的类型均为字符串。在前面的章节中，为了获取正确的产品价格，不得不将字符串解析成 BigDecimal 类型。为了便于复习，这里把第 4 章中 ProductController 类的 saveProduct 方法的部分代码复制过来了。

```
@RequestMapping(value="save-product")
public String saveProduct(ProductForm productForm, Model model) {
    logger.info("saveProduct called");
     // no need to create and instantiate a ProductForm
     // create Product
    Product product = new Product();
    product.setName(productForm.getName());
    product.setDescription(productForm.getDescription());
    try {
         product.setPrice(new BigDecimal(productForm.getPrice()));
    } catch (NumberFormatException e) {
    }
```

之所以需要解析 ProductForm 中的 price 属性，是因为它是一个 String，却需要用 BigDecimal 来填充 Product 的 price 属性。有了数据绑定，就可以用下面的代码取代上面的 saveProduct 方法部分。

```
@RequestMapping(value="save-product")
public String saveProduct(Product product, Model model)
```

有了数据绑定，就不再需要 ProductForm 类，也不需要解析 Product 对象的 price 属性了。

数据绑定的另一个好处是：当输入验证失败时，它会重新生成一个 HTML 表单。手工编写 HTML 代码时，必须记着用户之前输入的值，重新填充输入字段。有了 Spring 的数据绑定和表单标签库后，它们就会替你完成这些工作。

5.2 表单标签库

表单标签库中包含了可以用在 JSP 页面中渲染 HTML 元素的标签。为了使用这些标签，必须在 JSP 页面的开头处声明这个 taglib 指令。

```
<%@taglib prefix="form"
    uri="http://www.springframework.org/tags/form" %>
```

表 5.1 展示了表单标签库中的标签。

在接下来的小节中，将逐一介绍这些标签。5.3 节展示了一个范例应用程序，展示了数据绑定结合表单标签库的使用方法。

表 5.1　表单标签库中的标签

标签	描述
form	渲染表单元素
input	渲染<input type="text"/>元素
password	渲染<input type="password"/>元素
hidden	渲染<input type="hidden"/>元素
textarea	渲染 textarea 元素
checkbox	渲染一个<input type="checkbox"/>元素
checkboxes	渲染多个<input type="checkbox"/>元素
radiobutton	渲染一个<input type="radio"/>元素
radiobuttons	渲染多个<input type="radio"/>元素
select	渲染一个选择元素
option	渲染一个可选元素
options	渲染一个可选元素列表
errors	在 span 元素中渲染字段错误

5.2.1　表单标签

表单标签用于渲染 HTML 表单。要使用渲染一个表单输入字段的任何其他标签，必须有一个 form 标签。表单标签的属性见表 5.2。

表 5.2 中的所有标签都是可选的。这个表中没有包含 HTML 属性，如 method 和 action。

commandName 属性或许是其中最重要的属性，因为它定义了模型属性的名称，其中包含了一个表单支持对象（form backing object），其属性将用于填充所生成的表单。如果该属性存在，则必须在返回包含该表单的视图的请求处理方法中添加相应的模型属性。例如，在本章配套的 tags-demo 应用程序中，下列表单标签是在 BookAddForm.jsp 中定义的。

```
<form:form commandName="book" action="save-book" method="post">
    ...
</form:form>
```

表 5.2　表单标签的属性

属性	描述
acceptCharset	定义服务器接受的字符编码列表
commandName	暴露表单对象之模型属性的名称，默认为 command
cssClass	定义要应用到被渲染 form 元素的 CSS 类
cssStyle	定义要应用到被渲染 form 元素的 CSS 样式
htmlEscape	接受 true 或者 false，表示被渲染的值是否应该进行 HTML 转义
modelAttribute	暴露表单支持对象的模型属性名称，默认为 command

BookController 类中的 inputBook 方法，是返回 BookAddForm.jsp 的请求处理方法。下面就是 inputBook 方法。

```
@RequestMapping(value = "/input-book")
public String inputBook(Model model) {
    ...
    model.addAttribute("book", new Book());
    return "BookAddForm";
}
```

此处用 book 属性创建了一个 Book 对象，并添加到 Model。如果没有 Model 属性，BookAddForm.jsp 页面就会抛出异常，因为表单标签无法找到在其 commandName 属性中指定的 form backing object。

此外，一般来说，仍然需要使用 action 和 method 属性。这两个属性都是 HTML 属性，因此不在表 5.2 之中。

5.2.2　input 标签

input 标签渲染<input type="text"/>元素。这个标签最重要的属性是 path，它将这个输入字段绑定到表单支持对象的一个属性。例如，若随附<form/>标签的 commandName 属性值为 book，并且 input 标签的 path 属性值为 isbn，那么，input 标签将被绑定到 Book 对象的 isbn 属性。

表 5.3 展示了 input 标签的属性。表 5.3 中的属性都是可选的，其中不包含 HTML 属性。

表 5.3　input 标签的属性

属性	描述
cssClass	定义要应用到被渲染 input 元素的 CSS 类
cssStyle	定义要应用到被渲染 input 元素的 CSS 样式
cssErrorClass	定义要应用到被渲染 input 元素的 CSS 类，如果 bound 属性中包含错误，则覆盖 cssClass 属性值
htmlEscape	接受 true 或者 false，表示是否应该对被渲染的值进行 HTML 转义
path	要绑定的属性路径

举个例子，下面这个 input 标签被绑定到表单支持对象的 isbn 属性。

```
<form:input id="isbn" path="isbn" cssErrorClass="errorBox"/>
```

它将会被渲染成下面的<input/>元素：

```
<input type="text" id="isbn" name="isbn"/>
```

cssErrorClass 属性不起作用，除非 isbn 属性中有输入验证错误，并且采用同一个表单重新显示用户输入，在这种情况下，input 标签就会被渲染成下面这个 input 元素。

```
<input type="text" id="isbn" name="isbn" class="errorBox"/>
```

input 标签也可以绑定到嵌套对象的属性。例如，下列的 input 标签绑定到表单支持对象的 category 属性的 id 属性。

```
<form:input path="category.id"/>
```

5.2.3　password 标签

password 标签渲染<input type="password"/>元素，其属性见表 5.4。password 标签与 input

标签相似，只不过它有一个 showPassword 属性。

表 5.4 password 标签的属性

属性	描述
cssClass	定义要应用到被渲染 input 元素的 CSS 类
cssStyle	定义要应用到被渲染 input 元素的 CSS 样式
cssErrorClass	定义要应用到被渲染 input 元素的 CSS 类，如果 bound 属性中包含错误，则覆盖 cssClass 属性值
htmlEscape	接受 true 或者 false，表示是否应该对被渲染的值进行 HTML 转义
path	要绑定的属性路径
showPassword	表示应该显示或遮盖密码，默认值为 false

表 5.4 中的所有属性都是可选的，这个表中不包含 HTML 属性。下面是一个 password 标签的例子。

```
<form:password id="pwd" path="password" cssClass="normal"/>
```

5.2.4 hidden 标签

hidden 标签渲染<input type="hidden"/>元素，其属性见表 5.5。hidden 标签与 input 标签相似，只不过它没有可视的外观，因此不支持 cssClass 和 cssStyle 属性。

表 5.5 hidden 标签的属性

属性	描述
htmlEscape	接受 true 或者 false，表示是否应该对被渲染的值进行 HTML 转义
path	要绑定的属性路径

表 5.5 中的所有属性都是可选的，其中不包含 HTML 属性。

下面是一个 hidden 标签的例子。

```
<form:hidden path="productId"/>
```

5.2.5 textarea 标签

textarea 标签渲染一个 HTML 的 textarea 元素。Textarea 实际上就是支持多行输入的一个 input 元素。textarea 标签的属性见表 5.6。表 5.6 中的所有属性都是可选的，其中不包含 HTML 属性。

表 5.6　textarea 标签的属性

属性	描述
cssClass	定义要应用到被渲染 input 元素的 CSS 类
cssStyle	定义要应用到被渲染 input 元素的 CSS 样式
cssErrorClass	定义要应用到被渲染 input 元素的 CSS 类，如果 bound 属性中包含错误，则覆盖 cssClass 属性值
htmlEscape	接受 true 或者 false，表示是否应该对被渲染的值进行 HTML 转义
path	要绑定的属性路径

例如，下面的 textarea 标签就是被绑定到表单支持对象的 note 属性。

```
<form:textarea path="note" tabindex="4" rows="5" cols="80"/>
```

5.2.6　checkbox 标签

checkbox 标签渲染<input type="checkbox"/>元素。checkbox 标签的属性见表 5.7。表 5.7 中的所有属性都是可选的，其中不包含 HTML 属性。

表 5.7　checkbox 标签的属性

属性	描述
cssClass	定义要应用到被渲染 input 元素的 CSS 类
cssStyle	定义要应用到被渲染 input 元素的 CSS 样式
cssErrorClass	定义要应用到被渲染 input 元素的 CSS 类，如果 bound 属性中包含错误，则覆盖 cssClass 属性值
htmlEscape	接受 true 或者 false，表示是否应该对被渲染的（多个）值进行 HTML 转义
label	要作为标签用于被渲染复选框的值
path	要绑定的属性路径

例如，下面的 checkbox 标签绑定到 outOfStock 属性：

```
<form:checkbox path="outOfStock" value="Out of Stock"/>
```

5.2.7　radiobutton 标签

radiobutton 标签渲染<input type="radio"/>元素。radiobutton 标签的属性见表 5.8。表 5.8 中的所有属性都是可选的，其中不包含 HTML 属性。

例如，下列的 radiobutton 标签绑定到 newsletter 属性。

```
Computing Now <form:radiobutton path="newsletter" value="Computing Now"/>
<br/>
Modern Health <form:radiobutton path="newsletter" value="Modern Health"/>
```

表 5.8　radiobutton 标签的属性

属性	描述
cssClass	定义要应用到被渲染 input 元素的 CSS 类
cssStyle	定义要应用到被渲染 input 元素的 CSS 样式
cssErrorClass	定义要应用到被渲染 input 元素的 CSS 类，如果 bound 属性中包含错误，则覆盖 cssClass 属性值
htmlEscape	接受 true 或者 false，表示是否应该对被渲染的（多个）值进行 HTML 转义
label	要作为标签用于被渲染复选框的值
path	要绑定的属性路径

5.2.8　checkboxes 标签

checkboxes 标签渲染多个<input type="checkbox"/>元素。checkboxes 标签的属性见表 5.9。表 5.9 中的属性都是可选的，其中不包含 HTML 属性。

例如，下面的 checkboxes 标签将 model 属性 categoryList 的内容渲染为复选框。checkboxes 标签允许进行多个选择。

```
<form:checkboxes path="category" items="${categoryList}"/>
```

表 5.9　checkboxes 标签的属性

属性	描述
cssClass	定义要应用到被渲染 input 元素的 CSS 类
cssStyle	定义要应用到被渲染 input 元素的 CSS 样式
cssErrorClass	定义要应用到被渲染 input 元素的 CSS 类，如果 bound 属性中包含错误，则覆盖 cssClass 属性值
delimiter	定义两个 input 元素之间的分隔符，默认没有分隔符
element	给每个被渲染的 input 元素都定义一个 HTML 元素，默认为“span”
htmlEscape	接受 true 或者 false，表示是否应该对被渲染的（多个）值进行 HTML 转义
items	用于生成 input 元素的对象的 Collection、Map 或者 Array
itemLabel	item 属性中定义的 Collection、Map 或者 Array 中的对象属性，为每个 input 元素提供标签
itemValue	item 属性中定义的 Collection、Map 或者 Array 中的对象属性，为每个 input 元素提供值
path	要绑定的属性路径

5.2.9　radiobuttons 标签

radiobuttons 标签渲染多个<input type="radio"/>元素。radiobuttons 标签的属性见表 5.10。

例如，下面的 radiobuttons 标签将 model 属性 categoryList 的内容渲染为单选按钮。每次只能选择一个单选按钮。

```
<form:radiobuttons path="category" items="${categoryList}"/>
```

表 5.10　radiobuttons 标签的属性

属性	描述
cssClass	定义要应用到被渲染 input 元素的 CSS 类
cssStyle	定义要应用到被渲染 input 元素的 CSS 样式
cssErrorClass	定义要应用到被渲染 input 元素的 CSS 类，如果 bound 属性中包含错误，则覆盖 cssClass 属性值
delimiter	定义两个 input 元素之间的分隔符，默认没有分隔符
element	给每一个被渲染的 input 元素都定义一个 HTML 元素，默认为“span”
htmlEscape	接受 true 或者 false，表示是否应该对被渲染的（多个）值进行 HTML 转义
items	用于生成 input 元素的对象的 Collection、Map 或者 Array
itemLabel	item 属性中定义的 Collection、Map 或者 Array 中的对象属性，为每个 input 元素提供标签
itemValue	item 属性中定义的 Collection、Map 或者 Array 中的对象属性，为每个 input 元素提供值
path	要绑定的属性路径

5.2.10　select 标签

select 标签渲染一个 HTML 的 select 元素。被渲染元素的选项可能来自赋予其 items 属性的一个 Collection、Map、Array，或者来自一个嵌套的 option 或者 options 标签。select 标签的属性见表 5.11。表 5.11 中的所有属性都是可选的，其中不包含 HTML 属性。

items 属性特别有用，因为它可以绑定到对象的 Collection、Map、Array，为 select 元素生成选项。

例如，下面的 select 标签绑定到表单支持对象的 category 属性的 id 属性。它的选项来自 model 属性 categories。每个选项的值均来自 categories collection/map/array 的 id 属性，它的标签来自 name 属性。

```
<form:select id="category" path="category.id"
```

```
items="${categories}" itemLabel="name"
itemValue="id"/>
```

表 5.11 select 标签的属性

属性	描述
cssClass	定义要应用到被渲染 input 元素的 CSS 类
cssStyle	定义要应用到被渲染 input 元素的 CSS 样式
cssErrorClass	定义要应用到被渲染 input 元素的 CSS 类，如果 bound 属性中包含错误，则覆盖 cssClass 属性值
htmlEscape	接受 true 或者 false，表示是否应该对被渲染的（多个）值进行 HTML 转义
items	用于生成 input 元素的对象的 Collection、Map 或者 Array
itemLabel	item 属性中定义的 Collection、Map 或者 Array 中的对象属性，为每个 input 元素提供标签
itemValue	item 属性中定义的 Collection、Map 或者 Array 中的对象属性，为每个 input 元素提供值
path	要绑定的属性路径

5.2.11 option 标签

option 标签渲染 select 元素中使用的一个 HTML 的 option 元素，其属性见表 5.12。表 5.12 中的所有属性都是可选的，其中不包含 HTML 属性。

例如，下面是一个 option 标签的范例。

表 5.12 option 标签的属性

属性	描述
cssClass	定义要应用到被渲染 input 元素的 CSS 类
cssStyle	定义要应用到被渲染 input 元素的 CSS 样式
cssErrorClass	定义要应用到被渲染 input 元素的 CSS 类，如果 bound 属性中包含错误，则覆盖 cssClass 属性值
htmlEscape	接受 true 或者 false，表示是否应该对被渲染的（多个）值进行 HTML 转义

```
<form:select id="category" path="category.id"
        items="${categories}" itemLabel="name"
        itemValue="id">
    <option value="0">-- Please select --</option>
</form:select>
```

这个代码片断是渲染一个 select 元素，其选项来自 model 属性 categories，以及 option 标签。

5.2.12　options 标签

options 标签生成一个 HTML 的 option 元素列表，其属性见表 5.13，其中不包含 HTML 属性。

tags-demo 应用程序展示了一个 options 标签的范例。

表 5.13　options 标签的属性

属性	描述
cssClass	定义要应用到被渲染 input 元素的 CSS 类
cssStyle	定义要应用到被渲染 input 元素的 CSS 样式
cssErrorClass	定义要应用到被渲染 input 元素的 CSS 类，如果 bound 属性中包含错误，则覆盖 cssClass 属性值
htmlEscape	接受 true 或者 false，表示是否应该对被渲染的（多个）值进行 HTML 转义
items	用于生成 input 元素的对象的 Collection、Map 或者 Array
itemLabel	item 属性中定义的 Collection、Map 或者 Array 中的对象属性，为每个 input 元素提供标签
itemValue	item 属性中定义的 Collection、Map 或者 Array 中的对象属性，为每个 input 元素提供值

5.2.13　errors 标签

errors 标签渲染一个或者多个 HTML 的 span 元素，每个 span 元素中都包含一个字段错误。这个标签可以用于显示一个特定的字段错误，或者所有字段错误。

errors 标签的属性见表 5.14。表 5.14 中的所有属性都是可选的，其中不包含可能在 HTML 的 span 元素中出现的 HTML 属性。

例如，下面这个 errors 标签显示了所有字段错误。

```
<form:errors path="*"/>
```

下面的 errors 标签显示了一个与表单支持对象的 author 属性相关的字段错误。

```
<form:errors path="author"/>
```

表 5.14　errors 标签的属性

属性	描述
cssClass	定义要应用到被渲染 input 元素的 CSS 类
cssStyle	定义要应用到被渲染 input 元素的 CSS 样式

续表

属性	描述
Delimiter	分隔多个错误消息的分隔符
element	定义一个包含错误消息的 HTML 元素
htmlEscape	接受 true 或者 false，表示是否应该对被渲染的（多个）值进行 HTML 转义
path	要绑定的错误对象路径

5.3 数据绑定范例

在表单标签库中利用标签进行数据绑定的例子，参见 tags-demo 应用程序。这个范例围绕 domain 类 Book 进行。这个类中有几个属性，包括一个类型为 Category 的 category 属性。Category 有 id 和 name 两个属性。

这个应用程序允许列出书目、添加新书，以及编辑书目。

5.3.1 目录结构

图 5.1 中展示了 tags-demo 的目录结构。

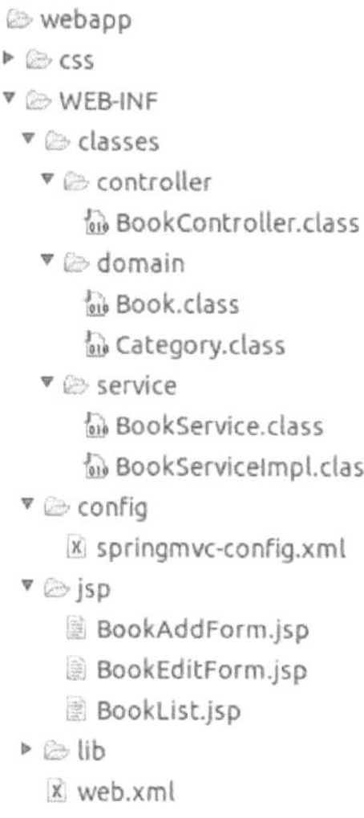

图 5.1 tags-demo 的目录结构

5.3.2 Domain 类

Book 类和 Category 类是这个应用程序中的 domain 类，它们分别如清单 5.1 和清单 5.2 所示。

清单 5.1　Book 类

```
package domain;
import java.math.BigDecimal;
import java.io.Serializable;

public class Book implements Serializable {

    private static final long serialVersionUID = 1L;
    private long id;
    private String isbn;
    private String title;
    private Category category;
    private String author;

    public Book() {
    }

    public Book(long id, String isbn, String title,
            Category category, String author, BigDecimal price) {
        this.id = id;
        this.isbn = isbn;
        this.title = title;
        this.category = category;
        this.author = author;
        this.price = price;
    }

    // get and set methods not shown

}
```

清单 5.2　Category 类

```
package domain;

import java.io.Serializable;

public class Category implements Serializable {
    private static final long serialVersionUID = 1L;
    private int id;
    private String name;

    public Category() {
    }

    public Category(int id, String name) {
```

```
        this.id = id;
        this.name = name;
    }

    // get and set methods not shown
}
```

5.3.3 Controller 类

下面的范例为 Book 提供了一个控制器：BookController 类。它允许用户创建新书目、更新书的详细信息，并在系统中列出所有书目。清单 5.3 中展示了 BookController 类。

清单 5.3 BookController 类

```
package controller;
import java.util.List;
import org.apache.commons.logging.Log;
import org.apache.commons.logging.LogFactory;
import org.springframework.beans.factory.annotation.Autowired;
import org.springframework.stereotype.Controller;
import org.springframework.ui.Model;
import org.springframework.web.bind.annotation.ModelAttribute;
import org.springframework.web.bind.annotation.PathVariable;
import org.springframework.web.bind.annotation.RequestMapping;
import domain.Book;
import domain.Category;
import service.BookService;

@Controller
public class BookController {

    @Autowired
    private BookService bookService;

    private static final Log logger =
            LogFactory.getLog(BookController.class);

    @RequestMapping(value = "/input-book ")
    public String inputBook(Model model) {
        List<Category> categories = bookService.getAllCategories();
        model.addAttribute("categories", categories);
        model.addAttribute("book", new Book());
        return "BookAddForm";
    }
    @RequestMapping(value = "/edit-book/{id}")
```

```
    public String editBook(Model model, @PathVariable long id) {
        List<Category> categories = bookService.getAllCategories();
        model.addAttribute("categories", categories);
        Book book = bookService.get(id);
        model.addAttribute("book", book);
        return "BookEditForm";
    }
    @RequestMapping(value = "/save-book")
    public String saveBook(@ModelAttribute Book book) {
        Category category =
                bookService.getCategory(book.getCategory().getId());
        book.setCategory(category);
        bookService.save(book);
        return "redirect:/list-book";
    }

    @RequestMapping(value = "/update-book")
    public String updateBook(@ModelAttribute Book book) {
        Category category =
                bookService.getCategory(book.getCategory().getId());
        book.setCategory(category);
        bookService.update(book);
        return "redirect:/list-book";
    }

    @RequestMapping(value = "/list-book")
    public String listBooks(Model model) {
        logger.info("listBooks");
        List<Book> books = bookService.getAllBooks();
        model.addAttribute("books", books);
        return "BookList";
    }
}
```

BookController 依赖 BookService 进行一些后台处理。@Autowired 注解用于给 BookController 注入一个 BookService 实现。

```
@Autowired
private BookService bookService;
```

5.3.4　Service 类

清单 5.4 和清单 5.5 分别展示了 BookService 接口和 BookServiceImpl 类。顾名思义，BookServiceImpl 就是实现 BookService。

清单 5.4 BookService 接口

```
package service;
import java.util.List;
import domain.Book;
import domain.Category;

public interface BookService {
    List<Category> getAllCategories();
    Category getCategory(int id);
    List<Book> getAllBooks();
    Book save(Book book);
    Book update(Book book);
    Book get(long id);
    long getNextId();
}
```

清单 5.5 BookServiceImpl 类

```
package service;
import java.util.ArrayList;
import java.util.List;
import org.springframework.stereotype.Service;
import domain.Book;
import domain.Category;

@Service
public class BookServiceImpl implements BookService {

    /*
     * this implementation is not thread-safe
     */
    private List<Category> categories;
    private List<Book> books;

    public BookServiceImpl() {
        categories = new ArrayList<Category>();
        Category category1 = new Category(1, "Computer");
        Category category2 = new Category(2, "Travel");
        Category category3 = new Category(3, "Health");
        categories.add(category1);
        categories.add(category2);
```

```
        categories.add(category3);

        books = new ArrayList<Book>();
        books.add(new Book(1L, "9781771970273",
                "Servlet & JSP: A Tutorial (2nd Edition)",
                category1, "Budi Kurniawan", new BigDecimal("54.99")));
        books.add(new Book(2L, "9781771970297",
                "C#: A Beginner's Tutorial (2nd Edition) ",
                category1, "Jayden Ky", new BigDecimal("39.99")));
    }

    @Override
    public List<Category> getAllCategories() {
        return categories;
    }

    @Override
    public Category getCategory(int id) {
        for (Category category : categories) {
            if (id == category.getId()) {
                return category;
            }
        }
        return null;
    }

    @Override
    public List<Book> getAllBooks() {
        return books;
    }

    @Override
    public Book save(Book book) {
        book.setId(getNextId());
        books.add(book);
        return book;
    }

    @Override
    public Book get(long id) {
        for (Book book : books) {
            if (id == book.getId()) {
                return book;
            }
        }
        return null;
```

```
    }

    @Override
    public Book update(Book book) {
        int bookCount = books.size();
        for (int i = 0; i < bookCount; i++) {
            Book savedBook = books.get(i);
            if (savedBook.getId() == book.getId()) {
                books.set(i, book);
                return book;
            }
        }
        return book;
    }

    @Override
    public long getNextId() {
        // needs to be locked
        long id = 0L;
        for (Book book : books) {
            long bookId = book.getId();
            if (bookId > id) {
                id = bookId;
            }
        }
        return id + 1;
    }
}
```

BookServiceImpl 类中包含了一个 Book 对象的 List 和一个 Category 对象的 List。这两个 List 都是在实例化类时生成的。这个类中还包含了获取所有书目、获取单个书目，以及添加和更新书目的方法。

5.3.5 配置文件

清单 5.6 展示了 tags-demo 中的 Spring MVC 配置文件。

清单 5.6 Spring MVC 配置文件

```
<?xml version="1.0" encoding="UTF-8"?>
<beans xmlns="http://www.springframework.org/schema/beans"
    xmlns:xsi="http://www.w3.org/2001/XMLSchema-instance"
    xmlns:p="http://www.springframework.org/schema/p"
    xmlns:mvc="http://www.springframework.org/schema/mvc"
    xmlns:context="http://www.springframework.org/schema/context"
```

```
    xsi:schemaLocation="
       http://www.springframework.org/schema/beans
       http://www.springframework.org/schema/beans/spring-beans.xsd
       http://www.springframework.org/schema/mvc
       http://www.springframework.org/schema/mvc/spring-mvc.xsd
       http://www.springframework.org/schema/context
       http://www.springframework.org/schema/context/spring-context.
       xsd">

    <context:component-scan base-package="controller"/>
    <context:component-scan base-package="service"/>

    ... <!-- other elements are not shown -->

</beans>
```

component-scan bean 使得 app05a.controller 包和 app05a.service 包得以被扫描。

5.3.6　视图

tags-demo 中使用的 3 个 JSP 页面如清单 5.7、清单 5.8 和清单 5.9 所示。BookAddForm.jsp 和 BookEditForm.jsp 页面中使用的是来自表单标签库的标签。

清单 5.7　BookList.jsp 页面

```
<%@ taglib uri="http://java.sun.com/jsp/jstl/core" prefix="c" %>
<!DOCTYPE HTML>
<html>
<head>
<title>Book List</title>
<style type="text/css">@import url("<c:url value="/css/main.css"/>");
</style>
</head>
<body>

<div id="global">
<h1>Book List</h1>
<a href="<c:url value="/book_input"/>">Add Book</a>
<table>
<tr>
    <th>Category</th>
    <th>Title</th>
    <th>ISBN</th>
    <th>Author</th>
    <th>Price</th>
    <th> </th>
```

```
</tr>
<c:forEach items="${books}" var="book">
    <tr>
        <td>${book.category.name}</td>
        <td>${book.title}</td>
        <td>${book.isbn}</td>
        <td>${book.author}</td>
        <td>${book.price}</td>
        <td><a href="book_edit/${book.id}">Edit</a></td>
    </tr>
</c:forEach>
</table>
</div>
</body>
</html>
```

清单 5.8 BookAddForm.jsp 页面

```
<%@ taglib prefix="form"
        uri="http://www.springframework.org/tags/form" %>
<%@ taglib uri="http://java.sun.com/jsp/jstl/core" prefix="c" %>
<!DOCTYPE HTML>
<html>
<head>
<title>Add Book Form</title>
<style type="text/css">@import url("<c:url
       value="/css/main.css"/>");</style>
</head>
<body>
<div id="global">

<form:form commandName="book" action="save-book" method="post">
    <fieldset>
        <legend>Add a book</legend>
        <p>
            <label for="category">Category: </label>
             <form:select id="category" path="category.id"
                items="${categories}" itemLabel="name"
                itemValue="id"/>
        </p>
        <p>
            <label for="title">Title: </label>
            <form:input id="title" path="title"/>
        </p>
        <p>
            <label for="author">Author: </label>
            <form:input id="author" path="author"/>
        </p>
```

```
            <p>
                <label for="isbn">ISBN: </label>
                <form:input id="isbn" path="isbn"/>
            </p>

            <p id="buttons">
                <input id="reset" type="reset" tabindex="4">
                <input id="submit" type="submit" tabindex="5"
                    value="Add Book">
            </p>
        </fieldset>
</form:form>
</div>
</body>
</html>
```

清单 5.9　BookEditForm.jsp 页面

```
<%@ taglib prefix="form"
       uri="http://www.springframework.org/tags/form" %>
<%@ taglib uri="http://java.sun.com/jsp/jstl/core" prefix="c" %>
<!DOCTYPE HTML>
<html>
<head>
<title>Edit Book Form</title>
<style type="text/css">@import url("<c:url
       value="/css/main.css"/>");</style>
</head>
<body>

<div id="global">
<c:url var="formAction" value="/update-book" />
<form:form commandName="book" action="${formAction}" method="post">
    <fieldset>
        <legend>Edit a book</legend>
        <form:hidden path="id"/>
        <p>
            <label for="category">Category: </label>
             <form:select id="category" path="category.id" items="$
       {categories}"
                itemLabel="name" itemValue="id"/>
        </p>
        <p>
            <label for="title">Title: </label>
            <form:input id="title" path="title"/>
        </p>
        <p>
            <label for="author">Author: </label>
```

```
            <form:input id="author" path="author"/>
        </p>
        <p>
            <label for="isbn">ISBN: </label>
            <form:input id="isbn" path="isbn"/>
        </p>

        <p id="buttons">
            <input id="reset" type="reset" tabindex="4">
            <input id="submit" type="submit" tabindex="5"
                value="Update Book">
        </p>
    </fieldset>
</form:form>
</div>
</body>
</html>
```

注意，在 BookEditForm.jsp 页面中，表单的 action 属性为<c:url />的值：

```
<c:url var="formAction" value ="/update-book"/>
<form:form commandName="book" action="${formAction}" method="post">
```

这是因为表单需要定位/update-book 映射，并且给 action 属性一个静态值 "update-book" 是有问题的。如果图书 ID 作为请求参数发送到编辑图书页面，则页面 URL 将如下所示：

```
http://domain/context/edit-book?id=1
```

该表单将正确提交到 http://domain/context/update-book。

但是，如果图书 ID 作为路径变量发送，页面网址将如下所示：

```
http://domain/context/edit-book/id
```

并且表单将被提交到：

```
http://domain/context/edit-book/update-book
```

因此，您应该使用<c:url/>来确保表单目标始终正确，无论网页网址如何。遗憾的是，表单的 action 属性不能取<c:url/>。因此，您需要创建一个变量 formAction 并从 action 属性引用它。

5.3.7 测试应用

要想测试这个应用程序范例，请打开以下网页：

```
http://localhost:8080/tags-demo/list-books
```

图 5.2 展示了第一次启动这个应用程序时显示的书目列表。

单击“Add Book”链接添加书目，或者单击书籍详情右侧的“Edit”链接来编辑书目。

图 5.3 展示了“Add a book”表单。图 5.4 中展示了“Edit a book”表单。

图 5.2　书目列表

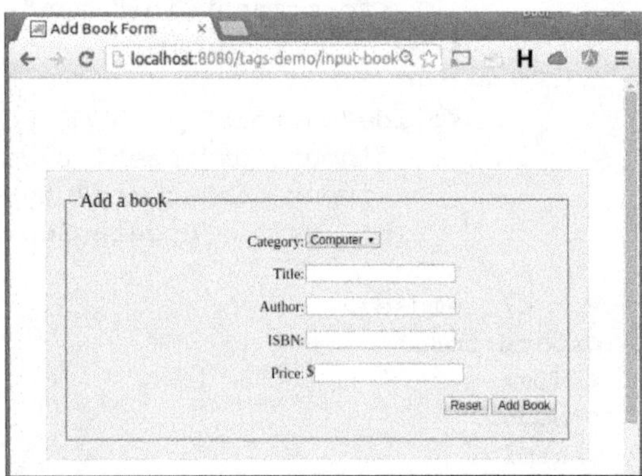

图 5.3　Add a book 表单

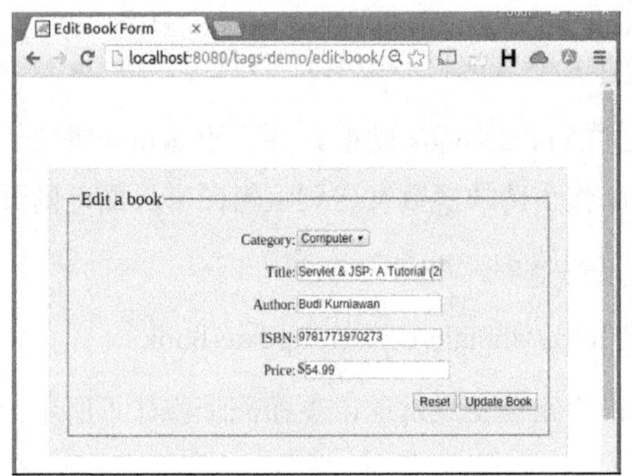

图 5.4　Edit a book 表单

5.4　小结

本章介绍了数据绑定和表单标签库中的标签。第 6 章和第 7 章将讨论如何进一步将数据绑定与 Converter、Formatter 以及验证器结合起来使用。

第 6 章 转换器和格式化

第 5 章已经见证了数据绑定的威力，并学习了如何使用表单标签库中的标签。但是，Spring 的数据绑定并非没有任何限制。有案例表明，Spring 在如何正确绑定数据方面是杂乱无章的。例如，Spring 总是试图用默认的语言区域将日期输入绑定到 java.util.Date。假如想让 Spring 使用不同的日期样式，就需要用一个 Converter（转换器）或者 Formatter（格式化）来协助 Spring 完成。

本章着重讨论 Converter 和 Formatter 的内容。这两者均可用于将一种对象类型转换成另一种对象类型。Converter 是通用元件，可以在应用程序的任意层中使用，而 Formatter 则是专门为 Web 层设计的。

本章有两个示例程序：converter-demo 和 formatter-demo。两者都使用一个 messageSource bean 来帮助显示受控的错误消息，这个 bean 的功能在第 10 章中介绍。

6.1 Converter

Spring 的 Converter 是可以将一种类型转换成另一种类型的一个对象。例如，用户输入的日期可能有许多种形式，如“December 25，2014”“12/25/2014”和“2014-12-25”，这些都表示同一个日期。默认情况下，Spring 会期待用户输入的日期样式与当前语言区域的日期样式相同。例如，对于美国的用户而言，就是月/日/年格式。如果希望 Spring 在将输入的日期字符串绑定到 LocalDate 时，使用不同的日期样式，则需要编写一个 Converter，才能将字符串转换成日期。java.time.LocalDate 类是 Java 8 的一个新类型，用来替代 java.util.Date。还需使用新的 Date/Time API 来替换旧有的 Date 和 Calendar 类。

为了创建 Converter，必须编写实现 org.springframework.core.convert.converter. Converter 接口的一个 Java 类。这个接口的声明如下：

```
public interface Converter<S, T>
```

这里的 S 表示源类型，T 表示目标类型。例如，为了创建一个可以将 Long 转换成 Date 的 Converter，要像下面这样声明 Converter 类：

```
public class MyConverter implements Converter<Long, LocalDate> {
}
```

在类实体中，需要编写一个来自 Converter 接口的 convert 方法实现。这个方法的签名如下：

```
T convert(S source)
```

例如，清单 6.1 展示了一个适用于任意日期样式的 Converter。

清单 6.1　String To LocalDate Converter

```
package converter;
import java.time.LocalDate;
import java.time.format.DateTimeFormatter;
import java.time.format.DateTimeParseException;
import org.springframework.core.convert.converter.Converter;

public class StringToLocalDateConverter implements Converter<String,
        LocalDate> {

    private String datePattern;

    public StringToLocalDateConverter(String datePattern) {
        this.datePattern = datePattern;
    }

    @Override
    public LocalDate convert(String s) {
        try {
            return LocalDate.parse(s, DateTimeFormatter.ofPattern(
                    datePattern));
        } catch (DateTimeParseException e) {
            // the error message will be displayed in <form:errors>
            throw new IllegalArgumentException(
                    "invalid date format. Please use this pattern\""
                            + datePattern + "\"");
        }
    }
}
```

注意清单 6.1 中的 Converter 方法，它利用传给构造器的日期样式，将一个 String 转换成 LocalDate。

为了使用 Spring MVC 应用程序中定制的 Converter，需要在 Spring MVC 配置文件中编写一个名为 conversionService 的 bean。bean 的类名称必须为 org.springframework.context.support.ConversionServiceFactoryBean。这个 bean 必须包含一个 converters 属性，它将列出要在应用程序中使用的所有定制 Converter。例如，下面的 bean 声明在清单 6.1 中注册了 StringToDateConverter。

```
<bean id="conversionService" class="org.springframework.context.support.
➥ConversionServiceFactoryBean">
    <property name="converters">
        <list>
            <bean class="converter.StringToLocalDateConverter">
                <constructor-arg type="java.lang.String"
                        value="MM-dd-yyyy"/>
            </bean>
        </list>
    </property>
</bean>
```

随后，要给 annotation-driven 元素的 conversion-service 属性赋值 bean 名称（本例中是 conversionService），如下所示：

```
<mvc:annotation-driven
        conversion-service="conversionService"/>
```

converter-demo 是一个范例应用程序，它利用 StringToLocalDateConverter 将 String 转换成 Employee 对象的 birthDate 属性。Employee 类如清单 6.2 所示。

清单 6.2 Employee 类

```
package domain;
import java.io.Serializable;
import java.time.LocalDate;

public class Employee implements Serializable {
    private static final long serialVersionUID = -908L;
    private long id;
    private String firstName;
    private String lastName;
    private LocalDate birthDate;
    private int salaryLevel;
```

```
    // getters and setters not shown
}
```

清单 6.3 中的 EmployeeController 类是 domain 对象 Employee 的控制器。

清单 6.3　converter-demo 中的 EmployeeController 类

```
package controller;

import org.springframework.stereotype.Controller;
import org.springframework.ui.Model;
import org.springframework.validation.BindingResult;
import org.springframework.validation.FieldError;
import org.springframework.web.bind.annotation.ModelAttribute;
import org.springframework.web.bind.annotation.RequestMapping;
import domain.Employee;

@Controller
public class EmployeeController {

    @RequestMapping(value="/add-employee")
    public String inputEmployee(Model model) {
        model.addAttribute(new Employee());
        return "EmployeeForm";
    }

    @RequestMapping(value="/save-employee")
    public String saveEmployee(@ModelAttribute Employee employee,
            BindingResult bindingResult, Model model) {

        if (bindingResult.hasErrors()) {
            FieldError fieldError = bindingResult.getFieldError();
            return "EmployeeForm";
        }

        // save employee here

        model.addAttribute("employee", employee);
        return "EmployeeDetails";
    }
}
```

EmployeeController 类有 inputEmployee 和 saveEmployee 两个处理请求的方法。inputEmployee 方法返回清单 6.4 中的 EmployeeForm.jsp 页面。saveEmployee 方法取出在提交 Employee 表单时创建的一个 Employee 对象。有了 StringToLocalDateConverter converter，就

不需要劳驾控制器类将字符串转换成日期了。

saveEmployee 方法的 BindingResult 参数中放置了 Spring 的所有绑定错误。该方法利用 BindingResult 记录所有绑定错误。绑定错误也可以利用 errors 标签显示在一个表单中，如 EmployeeForm.jsp 页面所示。

清单 6.4 EmployeeForm.jsp 页面

```
<%@ taglib prefix="form"uri="http://www.springframework.org/tags/form" %>
<%@ taglib uri="http://java.sun.com/jsp/jstl/core" prefix="c" %>
<!DOCTYPE HTML>
<html>
<head>
<title>Add Employee Form</title>
<style type="text/css">@import url("<c:url
       value="/css/main.css"/>");</style>
</head>
<body>

<div id="global">
<form:form commandName="employee" action="save-employee" method="post">
    <fieldset>
        <legend>Add an employee</legend>
        <p>
            <label for="firstName">First Name: </label>
            <form:input path="firstName" tabindex="1"/>
        </p>
        <p>
            <label for="lastName">First Name: </label>
            <form:input path="lastName" tabindex="2"/>
        </p>
        <p>
            <form:errors path="birthDate" cssClass="error"/>
        </p>
        <p>
            <label for="birthDate">Date Of Birth: </label>
            <form:input path="birthDate" tabindex="3" />
        </p>
        <p id="buttons">
            <input id="reset" type="reset" tabindex="4">
            <input id="submit" type="submit" tabindex="5"
                value="Add Employee">
        </p>
    </fieldset>
</form:form>
```

```
</div>
</body>
</html>
```

在浏览器中打开以下 URL，可以对这个转换器进行测试：

```
http://localhost:8080/converter-demo/add-employee
```

试着输入一个无效的日期，将会被转到同一个 Employee 表单，并且可以在表单中看到错误消息，如图 6.1 所示。

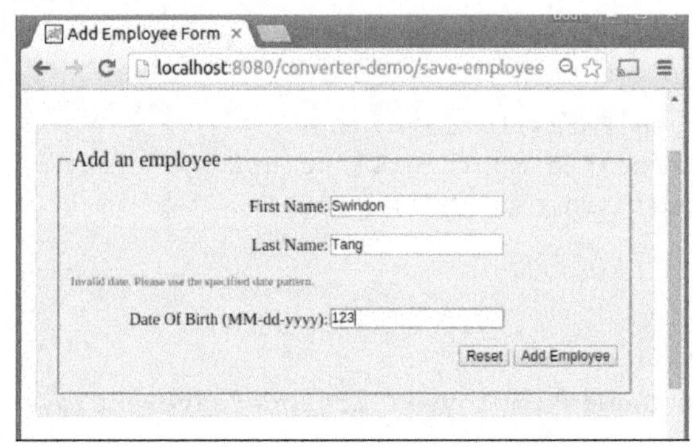

图 6.1　Employee 表单中的转换错误

6.2　Formatter

Formatter 就像 Converter 一样，也是将一种类型转换成另一种类型。但是，Formatter 的源类型必须是一个 String，而 Converter 则适用于任意的源类型。Formatter 更适合 Web 层，而 Converter 则可以用在任意层中。为了转换 Spring MVC 应用程序表单中的用户输入，始终应该选择 Formatter，而不是 Converter。

为了创建 Formatter，要编写一个实现 org.springframework.format.Formatter 接口的 Java 类。下面是这个接口的声明：

```
public interface Formatter<T>
```

这里的 T 表示输入字符串要转换的目标类型。该接口有 parse 和 print 两个方法，所有实现都必须覆盖它们。

```
T parse(String text, java.util.Locale locale)
String print(T object, java.util.Locale locale)
```

parse 方法利用指定的 Locale 将一个 String 解析成目标类型。print 方法与之相反，它返回目标对象的字符串表示法。

例如，formatter-demo 应用程序中用一个 LocalDateFormatter 将 String 转换成 Date。其作用与 converter-demo 中的 StringToLocalDateConverter 一样。

DateFormatter 类如清单 6.5 所示。

清单 6.5 Local DateFormatter 类

```
package formatter;
import java.text.ParseException;
import java.time.LocalDate;
import java.time.format.DateTimeFormatter;
import java.time.format.DateTimeParseException;
import java.util.Locale;
import org.springframework.format.Formatter;

public class LocalDateFormatter implements Formatter<LocalDate> {

    private DateTimeFormatter formatter;
    private String datePattern;

    public LocalDateFormatter(String datePattern) {
        this.datePattern = datePattern;
        formatter= DateTimeFormatter.ofPattern(datePattern);
    }

    @Override
    public String print(LocalDate date, Locale locale) {
        return date.format(formatter);
    }

    @Override
    public LocalDate parse(String s, Locale locale)
            throws ParseException {
         try {
            return LocalDate.parse(s,
                    DateTimeFormatter.ofPattern(datePattern));
        } catch (DateTimeParseException e) {
            // the error message will be displayed in <form:errors>
            throw new IllegalArgumentException(
                    "invalid date format. Please use this pattern\""
                            + datePattern + "\"");
```

```
            }
        }
    }
```

为了在 Spring MVC 应用程序中使用 Formatter，需要利用名为 conversionService 的 bean 对它进行注册。bean 的类名称必须为 org.springframework.format.support.FormattingConversionServiceFactoryBean。这与 converter-demo 中用于注册 converter 的类不同。这个 bean 可以用一个 formatters 属性注册 formatter，用一个 converters 属性注册 converter。清单 6.6 展示了 formatter-demo 的 Spring 配置文件。

清单 6.6　formatter-demo 的 Spring 配置文件

```
<?xml version="1.0" encoding="UTF-8"?>
<beans xmlns="http://www.springframework.org/schema/beans"
    xmlns:xsi="http://www.w3.org/2001/XMLSchema-instance"
    xmlns:p="http://www.springframework.org/schema/p"
    xmlns:mvc="http://www.springframework.org/schema/mvc"
    xmlns:context="http://www.springframework.org/schema/context"
    xsi:schemaLocation="
       http://www.springframework.org/schema/beans
       http://www.springframework.org/schema/beans/spring-beans.xsd
       http://www.springframework.org/schema/mvc
       http://www.springframework.org/schema/mvc/spring-mvc.xsd
       http://www.springframework.org/schema/context
       http://www.springframework.org/schema/context/spring-context.xsd">

    <context:component-scan base-package="controller"/>
    <context:component-scan base-package="formatter"/>

    <mvc:annotation-driven conversion-service="conversionService"/>

    <mvc:resources mapping="/css/**" location="/css/"/>
    <mvc:resources mapping="/*.html" location="/"/>

    <bean id="viewResolver"
          class="org.springframework.web.servlet.view.
➥InternalResourceViewResolver">
        <property name="prefix" value="/WEB-INF/jsp/" />
        <property name="suffix" value=".jsp" />
    </bean>

    <bean id="conversionService"
          class="org.springframework.format.support.
➥FormattingConversionServiceFactoryBean">
       <property name="formatters">
```

```
            <set>
                <bean class="formatter.LocalDateFormatter">
                    <constructor-arg type="java.lang.String"
                       value="MM-dd-yyyy" />
                </bean>
            </set>
        </property>
    </bean>
</beans>
```

注意，还需要给这个 Formatter 添加一个 component-scan 元素。

在浏览器中打开下面的 URL，可以测试 app06b 中的 Formatter：

```
http://localhost:8080/formatter-demo/add-employee
```

6.3 用 Registrar 注册 Formatter

注册 Formatter 的另一种方法是使用 Registrar。例如，清单 6.7 就是注册 DateFormatter 的一个例子。

清单 6.7 MyFormatterRegistrar 类

```
package formatter;
import org.springframework.format.FormatterRegistrar;
import org.springframework.format.FormatterRegistry;

public class MyFormatterRegistrar implements FormatterRegistrar {

    private String datePattern;
    public MyFormatterRegistrar(String datePattern) {
        this.datePattern = datePattern;
    }

    @Override
    public void registerFormatters(FormatterRegistry registry) {
        registry.addFormatter(new LocalDateFormatter(datePattern));
        // register more formatters here
    }
}
```

有了 Registrar，就不需要在 Spring MVC 配置文件中注册任何 Formatter 了，只在 Spring 配置文件中注册 Registrar 就可以了，如清单 6.8 所示。

清单 6.8　在 springmvc-config.xml 文件中注册 Registrar

```
<?xml version="1.0" encoding="UTF-8"?>
<beans xmlns="http://www.springframework.org/schema/beans"
    xmlns:xsi="http://www.w3.org/2001/XMLSchema-instance"
    xmlns:p="http://www.springframework.org/schema/p"
    xmlns:mvc="http://www.springframework.org/schema/mvc"
    xmlns:context="http://www.springframework.org/schema/context"
    xsi:schemaLocation="
        http://www.springframework.org/schema/beans
        http://www.springframework.org/schema/beans/spring-beans.xsd
        http://www.springframework.org/schema/mvc
        http://www.springframework.org/schema/mvc/spring-mvc.xsd
        http://www.springframework.org/schema/context
        http://www.springframework.org/schema/context/springcontext.xsd">

    <context:component-scan base-package="controller" />
    <context:component-scan base-package="formatter" />

    <mvc:annotation-driven conversion-service="conversionService" />

    <mvc:resources mapping="/css/**" location="/css/" />
    <mvc:resources mapping="/*.html" location="/" />

    <bean id="viewResolver"
            class="org.springframework.web.servlet.view.
➥InternalResourceViewResolver">
        <property name="prefix" value="/WEB-INF/jsp/" />
        <property name="suffix" value=".jsp" />
    </bean>

    <bean id="conversionService"
            class="org.springframework.format.support.
➥FormattingConversionServiceFactoryBean">

        <property name="formatterRegistrars">
            <set>
                <bean class="formatter.MyFormatterRegistrar">
                    <constructor-arg type="java.lang.String"
                            value="MM-dd-yyyy" />
                </bean>
            </set>
        </property>
    </bean>
</beans>
```

6.4 选择 Converter，还是 Formatter

Converter 是一般工具，可以将一种类型转换成另一种类型。例如，将 String 转换成 LocalDate，或者将 Long 转换成 LocalDate。Converter 既可以用在 Web 层，也可以用在其他层中。

Formatter 只能将 String 转换成另一种 Java 类型。例如，将 String 转换成 LocalDate，但它不能将 Long 转换成 LocalDate。因此，Formatter 适用于 Web 层。为此，在 Spring MVC 应用程序中，选择 Formatter 比选择 Converter 更合适。

6.5 小结

本章学习了 Converter 和 Formatter，可以利用它们来引导 Spring MVC 应用程序中的数据绑定。Converter 是一般工具，可以将任意类型转换成另一种类型，而 Formatter 则只能将 String 转换成另一种 Java 类型。Formatter 更适用于 Web 层。

第 7 章 验证器

输入验证是 Spring 处理的最重要 Web 开发任务之一。在 Spring MVC 中，有两种方式可以验证输入，即利用 Spring 自带的验证框架，或者利用 JSR 303 实现。本章将详细介绍这两种输入验证方法。

本章用两个不同的示例分别介绍两种方式：spring-validator 和 jsr303-validator。

7.1 验证概览

Converter 和 Formatter 作用于字段级。在 MVC 应用程序中，它们将 String 转换或格式化成另一种 Java 类型，如 java.time.LocalDate。验证器则作用于对象级。它决定某一个对象中的所有字段是否均是有效的，以及是否遵循某些规则。一个典型的 Spring MVC 应用会同时应用到 formatters/converters 和 validators。

如果一个应用程序中既使用了 Formatter，又有 validator（验证器），那么，应用中的事件顺序是这样的：在调用 Controller 期间，将会有一个或者多个 Formatter，试图将输入字符串转换成 domain 对象中的 field 值。一旦格式化成功，验证器就会介入。

例如，Order 对象可能会有一个 shippingDate 属性（其类型显然为 LocalDate），它的值绝对不可能早于今天的日期。当调用 OrderController 时，DateFormatter 会将字符串转化成 Date，并将它赋予 Order 对象的 shippingDate 属性。如果转换失败，用户就会被转回到前一个表单。如果转换成功，则会调用验证器，查看 shippingDate 是否早于今天的日期。

现在，你或许会问，将验证逻辑转移到 LocalDateFormatter 中是否更加明智？因为比较一下日期并非难事，但答案却是否定的。首先，LocalDateFormatter 还可用于将其他字符串格式化成日期，如 birthDate 或者 purchaseDate。这两个日期的规则都不同于 shippingDate。事实上，比如，员工的出生日期绝对不可能晚于今日。其次，验证器可以检验两个或更多字段之间的关系，各字段均受不同的 Formatter 支持。例如，假设 Employee 对象有 birthDate 属性和 startDate 属性，验证器就可以设定规则，使任何员工的入职日期均不可能早于他的出生日期。因此，有效的 Employee 对象必须让它的 birthDate 属性值早于其 startDate 值。这就是验证器的任务。

7.2 Spring 验证器

从一开始，Spring 就设计了输入验证，甚至早于 JSR 303（Java 验证规范）。因此，Spring 的 Validation 框架至今都很普遍，尽管对于新项目，一般建议使用 JSR 303 验证器。

为了创建 Spring 验证器，要实现 org.springframework.validation.Validator 接口。这个接口如清单 7.1 所示，其中有 supports 和 validate 两个方法。

清单 7.1 Spring 的 Validator 接口

```
package org.springframework.validation;
public interface Validator {
    boolean supports(Class<?> clazz);
    void validate(Object target, Errors errors);
}
```

如果验证器可以处理指定的 Class，supports 方法将返回 true。validate 方法会验证目标对象，并将验证错误填入 Errors 对象。

Errors 对象是 org.springframework.validation.Errors 接口的一个实例。Errors 对象中包含了一系列 FieldError 和 ObjectError 对象。FieldError 表示与被验证对象中的某个属性相关的一个错误。例如，如果产品的 price 属性必须为负数，并且 Product 对象被验证为负数，那么就需要创建一个 FieldError。例如，在欧洲出售的一本 Book，却在美国的网店上购买，那么就会出现一个 ObjectError。

编写验证器时，不需要直接创建 Error 对象，因为实例化 ObjectError 或 FieldError 花费了大量的编程精力。这是因为 ObjectError 类的构造器需要 4 个参数，FieldError 类的构造器则需要 7 个参数，如以下构造器签名所示：

```
ObjectError(String objectName, String[] codes, Object[] arguments,
```

```
        String defaultMessage)

FieldError(String objectName, String field, Object rejectedValue,
        boolean bindingFailure, String[] codes, Object[] arguments,
        String defaultMessage)
```

给 Errors 对象添加错误的最容易的方法是：在 Errors 对象上调用一个 reject 或者 rejectValue 方法。调用 reject，会往 FieldError 中添加一个 ObjectError 和 rejectValue。

下面是 reject 和 rejectValue 的部分方法重载：

```
void reject(String errorCode)
void reject(String errorCode, String defaultMessage)
void rejectValue(String field, String errorCode)
void rejectValue(String field, String errorCode,
        String defaultMessage)
```

大多数时候，只给 reject 或者 rejectValue 方法传入一个错误码，Spring 就会在属性文件中查找错误码，获得相应的错误消息。还可以传入一个默认消息，当没有找到指定的错误码时，就会使用默认消息。

Errors 对象中的错误消息，可以利用表单标签库的 Errors 标签显示在 HTML 页面中。错误消息可以通过 Spring 支持的国际化特性本地化。关于国际化的更多信息，请查看第 10 章。

7.3　ValidationUtils 类

org.springframework.validation.ValidationUtils 类是一个工具，有助于编写 Spring 验证器。不需要像下面这样编写：

```
if (firstName == null || firstName.isEmpty()) {
    errors.rejectValue("price");
}
```

而是可以利用 ValidationUtils 类的 rejectIfEmpty 方法，像下面这样：

```
ValidationUtils.rejectIfEmpty("price");
```

或者下面这样的代码：

```
if (firstName == null || firstName.trim().isEmpty()) {
    errors.rejectValue("price");
}
```

可以编写成：

```
ValidationUtils.rejectIfEmptyOrWhitespace("price");
```

下面是 validationUtils 中 rejectIfEmpty 和 rejectIfEmptyOrWhitespace 方法的方法重载：

```
public static void rejectIfEmpty(Errors errors, String field,
        String errorCode)

public static void rejectIfEmpty(Errors errors, String field,
        String errorCode, Object[] errorArgs)

public static void rejectIfEmpty(Errors errors, String field,
        String errorCode, Object[] errorArgs, String defaultMessage)

public static void rejectIfEmpty(Errors errors, String field,
        String errorCode, String defaultMessage)

public static void rejectIfEmptyOrWhitespace(Errors errors,
        String field, String errorCode)

public static void rejectIfEmptyOrWhitespace(Errors errors,
        String field, String errorCode, Object[] errorArgs)

public static void rejectIfEmptyOrWhitespace(Errors errors,
        String field, String errorCode, Object[] errorArgs,
        String defaultMessage)

public static void rejectIfEmptyOrWhitespace(Errors errors,
        String field, String errorCode, String defaultMessage)
```

此外，ValidationUtils 还有一个 invokeValidator 方法，用来调用验证器。

```
public static void invokeValidator(Validator validator,
        Object obj, Errors errors)
```

接下来的小节将通过范例来介绍如何使用这个工具。

7.4 Spring 的 Validator 范例

spring-validator 应用程序中包含一个名为 ProductValidator 的验证器，用于验证 Product 对象。Spring-validator 的 Product 类如清单 7.2 所示。ProductValidator 类如清单 7.3 所示。

清单 7.2 Product 类

```
package domain;
import java.io.Serializable;
```

```
import java.math,BigDecimal;
import java.time.LocalDate;

public class Product implements Serializable {
    private static final long serialVersionUID = 1L;
    private String name;
    private String description;
    private BigDecimal price;
    private LocalDate productionDate;

    //getters and setters methods not shown
}
```

清单 7.3　ProductValidator 类

```
package validator;
import java.math,BigDecimal;
import java.time.LocalDate;
import org.springframework.validation.Errors;
import org.springframework.validation.ValidationUtils;
import org.springframework.validation.Validator;
import domain.Product;

public class ProductValidator implements Validator {
    @Override
    public boolean supports(Class<?> klass) {
        return Product.class.isAssignableFrom(klass);
    }

    @Override
    public void validate(Object target, Errors errors) {
        Product product = (Product) target;
        ValidationUtils.rejectIfEmpty(errors, "name",
                "productname.required");
        ValidationUtils.rejectIfEmpty(errors, "price","price.required");
        ValidationUtils.rejectIfEmpty(errors, "productionDate",
                "productiondate.required");
        BigDecimal price = product.getPrice();
        if (price != null && price.compareTo(BigDecimal.ZERO) < 0) {
            errors.rejectValue("price", "price.negative");
        }
        Local Date productionDate = product.getProductionDate();
        if (productionDate != null) {
            if (productionDate.isAfter(LocalDate.now())) {
                errors.rejectValue("productionDate",
                        "productiondate.invalid");
```

```
                }
            }
        }
    }
```

ProductValidator 验证器是一个非常简单的验证器。它的 validate 方法会检验 Product 是否有名称和价格，并且价格是否不为负数。它还会确保生产日期不晚于今天。

7.5 源文件

验证器不需要显式注册，但是如果想要从某个属性文件中获取错误消息，则需要通过声明 messageSource bean，告诉 Spring 要去哪里查找这个文件。下面是 app07a 中的 messageSource bean：

```
<bean id="messageSource" class="org.springframework.context.support.
➥ReloadableResourceBundleMessageSource">
    <property name="basename" value="/WEB-INF/resource/messages"/>
</bean>
```

这个 bean 实际上是说，错误码和错误消息可以在/WEB-INF/resource 目录下的 messages.properties 文件中找到。

清单 7.4 展示了 messages.properties 文件的内容。

清单 7.4 messages.properties 文件

```
productname.required.product.name=Please enter a product name
price.required=Please enter a price
price.negative=Price cannot be less than 0
productiondate.required=Please enter a production date
productiondate.invalid=Please ensure the production date is not later
➥than today
typeMismatch.productionDate=Invalid production date
```

7.6 Controller 类

在 Controller 类中通过实例化 validator 类，可以使用 Spring 验证器。清单 7.5 中 ProductController 类的 saveProduct 方法创建了一个 ProductValidator，并调用其 validate 方法。为了检验该验证器是否生成错误消息，需在 BindingResult 中调用 hasErrors 方法。

清单 7.5 ProductController 类

```
package controller;
import org.apache.commons.logging.Log;
import org.apache.commons.logging.LogFactory;
import org.springframework.stereotype.Controller;
import org.springframework.ui.Model;
import org.springframework.validation.BindingResult;
import org.springframework.validation.FieldError;
import org.springframework.web.bind.annotation.ModelAttribute;
import org.springframework.web.bind.annotation.RequestMapping;
import domain.Product;
import validator.ProductValidator;

@Controller
public class ProductController {

    private static final Log logger = LogFactory
            .getLog(ProductController.class);

    @RequestMapping(value = "/add-product")
    public String inputProduct(Model model) {
        model.addAttribute("product", new Product());
        return "ProductForm";
    }
    @RequestMapping(value = "/save-product")
    public String saveProduct(@ModelAttribute Product product,
            BindingResult bindingResult, Model model) {
        ProductValidator productValidator = new ProductValidator();
        productValidator.validate(product, bindingResult);

        if (bindingResult.hasErrors()) {
            FieldError fieldError = bindingResult.getFieldError();
            logger.debug("Code:" + fieldError.getCode() + ", field:"

                    + fieldError.getField());

            return "ProductForm";
        }

        // save product here

        model.addAttribute("product", product);
        return "ProductDetails";
    }
}
```

使用 Spring 验证器的另一种方法是：在 Controller 中编写 initBinder 方法，并将验证器传到 WebDataBinder，并调用其 validate 方法。

```
@org.springframework.web.bind.annotation.InitBinder
public void initBinder(WebDataBinder binder) {
    // this will apply the validator to all request-handling methods
    binder.setValidator(new ProductValidator());
    binder.validate();
}
```

将验证器传到 WebDataBinder，会使该验证器应用于 Controller 类中所有处理请求的方法。

或者利用@javax.validation.Valid 对要验证的对象参数进行注解。例如：

```
public String saveProduct(@Valid @ModelAttribute Product product,
        BindingResult bindingResult, Model model) {
```

Valid 注解类型是在 JSR 303 中定义的。关于 JSR 303 的相关信息，7.8 节将会讨论。

7.7　测试验证器

要想测试 spring-validator 中的验证器，在浏览器中打开以下 URL：

```
http://localhost:8080/spring-validator/add-product
```

你将会看到一张空白的 Product 表。如果单击“Add Product”按钮，没有输入任何值，将会被转回 Product 表，并且这次验证器会显示出错消息，如图 7.1 所示。

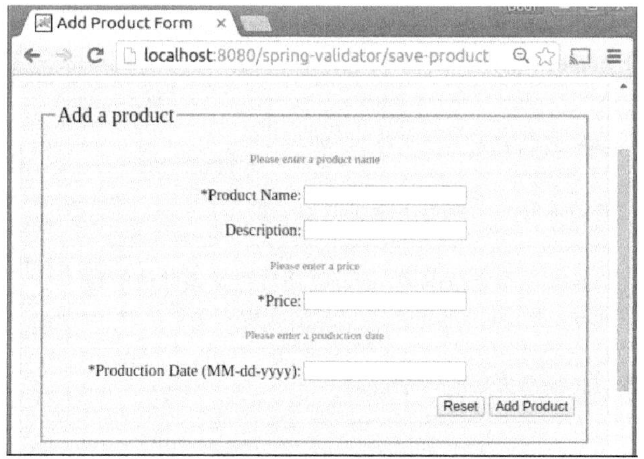

图 7.1　ProductValidator 的效果图

7.8　JSR 303 验证

JSR 303 “Bean Validation”（发布于 2009 年 11 月）和 JSR 349 “Bean Validation 1.1”（发布于 2013 年 5 月）指定了一整套 API，通过注解给对象属性添加约束。JSR 303 和 JSR 349 可以分别从以下网址下载：

```
http://jcp.org/en/jsr/detail?id=303
http://jcp.org/en/jsr/detail?id=349
```

当然，JSR 只是一个规范文档，本身用处不大，除非编写了它的实现。对于 JSR bean validation，目前有两个实现。第一个实现是 Hibernate Validator，目前版本为 5，JSR 303 和 JSR 349 它都实现了，可从以下网站下载：

```
http://sourceforge.net/projects/hibernate/files/hibernate-validator/
```

第二个实现是 Apache BVal，可从以下网站下载：

```
http://bval.apache.org/downloads.html
```

JSR 303 不需要编写验证器，但要利用 JSR 303 注解类型嵌入约束。JSR 303 约束见表 7.1。

表 7.1　JSR 303 约束

属性	描述	范例
@AssertFalse	应用于 boolean 属性，该属性值必须为 False	@AssertFalse boolean hasChildren;
@AssertTrue	应用于 boolean 属性，该属性值必须为 True	@AssertTrue boolean isEmpty;
@DecimalMax	该属性值必须为小于或等于指定值的小数	@DecimalMax("1.1") BigDecimal price;
@DecimalMin	该属性值必须为大于或等于指定值的小数	@DecimalMin("0.04") BigDecimal price;
@Digits	该属性值必须在指定范围内。integer 属性定义该数值的最大整数部分，fraction 属性定义该数值的最大小数部分	@Digits(integer=5, fraction=2) BigDecimal price;
@Future	该属性值必须是未来的一个日期	@Future Date shippingDate;
@Max	该属性值必须是一个小于或等于指定值的整数	@Max(150) int age;

续表

属性	描述	范例
@Min	该属性值必须是一个大于或等于指定值的整数	@Max(150) int age;
@NotNull	该属性值不能为 Null	@NotNull String firstName;
@Null	该属性值必须为 Null	@Null String testString;
@Past	该属性值必须是过去的一个日期	@Past Date birthDate;
@Pattern	该属性值必须与指定的常规表达式相匹配	@Pattern(regext="\\d{3 }") String areaCode;
@Size	该属性值必须在指定范围内	Size(min=2, max=140) String description;

一旦了解了 JSR 303 validation 的使用方法，使用起来会比 Spring 验证器还要容易。像使用 Spring 验证器一样，可以在属性文件中以下列格式来使用 property 键，覆盖来自 JSR 303 验证器的错误消息：

```
constraint.object.property
```

例如，为了覆盖以@Size 注解约束的 Product 对象的 name 属性，可以在属性文件中使用下面这个键：

```
Size.product.name
```

为了覆盖以@Past 注解约束的 Product 对象的 productionDate 属性，可以在属性文件中使用下面这个键：

```
Past.product.productionDate
```

7.9　JSR 303 Validator 范例

jsr303-validator 应用程序展示了 JSR 303 输入验证的例子。这个应用程序是对 spring-validator 进行修改之后的版本，与之前的版本有一些区别。首先，它没有 ProductValidator 类。其次，来自 Hibernate Validator 库的 jar 文件已经被添加到了 WEB-INF/lib 中。

清单 7.6 Product 类的 name 和 productionDate 字段已经用 JSR 303 注解类型进行了注解。

清单 7.6 Product 类

```
package domain;
import java.io.Serializable;
import java.math.BigDecimal
import java.time.LocalDate;
import javax.validation.constraints.Past;
import javax.validation.constraints.Size;

public class Product implements Serializable {
    private static final long serialVersionUID = 78L;

    @Size(min=1, max=10)
    private String name;

    private String description;
    private BigDecimal price;

    @Past
    private LocalDate productionDate;

    // getters and setters not shown
}
```

在 ProductController 类的 saveProduct 方法中，必须用@Valid 对 Product 参数进行注解，如清单 7.7 所示。

清单 7.7 ProductController 类

```
package controller;
import javax.validation.Valid;
import org.apache.commons.logging.Log;
import org.apache.commons.logging.LogFactory;
import org.springframework.stereotype.Controller;
import org.springframework.ui.Model;
import org.springframework.validation.BindingResult;
import org.springframework.validation.FieldError;
import org.springframework.web.bind.annotation.ModelAttribute;
import org.springframework.web.bind.annotation.RequestMapping;
import domain.Product;

@Controller
public class ProductController {

    private static final Log logger = LogFactory
            .getLog(ProductController.class);
```

```
    @RequestMapping(value = "/add-product")
    public String inputProduct(Model model) {
        model.addAttribute("product", new Product());
        return "ProductForm";
    }

    @RequestMapping(value = "/save-product")
    public String saveProduct(@Valid @ModelAttribute Product product,
            BindingResult bindingResult, Model model) {

        if (bindingResult.hasErrors()) {
            FieldError fieldError = bindingResult.getFieldError();
            logger.info("Code:" + fieldError.getCode() + ", object:"
                    + fieldError.getObjectName() + ", field:"
                    + fieldError.getField());
            return "ProductForm";
         }

         // save product here

         model.addAttribute("product", product);
         return "ProductDetails";
    }
}
```

为了定制来自验证器的错误消息，要在 messages.properties 文件中使用两个键。messages.properties 文件如清单 7.8 所示。

清单 7.8 jsr303-validator 中的 messages.properties 文件

```
productName.required.product.name=Please enter a product name
price.required=Please enter a price
price.negative=Price cannot be less than 0
productionDate.required=Please enter a production date
productionDate.invalid=Please ensure the production date is not later
➥than today
typeMismatch.productionDate=Invalid production date
Past.productionDate=Production date must be a past date
```

要想测试 jsr303-validator 中的验证器，可以在浏览器中打开以下网址：

```
http://localhost:8080/jsr303-validator/add-product
```

请注意，在编写本书时，最新版本（版本 5.2.4）的 Hibernate Validator 仍然无法验证使用@Past 或@Future 注解的 LocalDate 或 LocalDateTime 类型的字段。因为 LocalDate 是 Date / Time

API 中的新类型，应该用来代替 java.util.Date。 因此，这个项目包含一个自定义验证器和重写过的 Hibernate 验证器的 ConstraintHelper 类。

7.10　小结

本章学习了可以在 Spring MVC 应用程序中使用的两种验证器：Spring MVC 验证器和 JSR 303 验证器。由于 JSR 303 是正式的 Java 规范，因此建议新项目使用 JSR 303 验证器。

第 8 章 表达式语言

JSP 2.0 最重要的特性之一就是表达式语言（EL），JSP 用户可以用它来访问应用程序数据。由于受到 ECMAScript 和 XPath 表达式语言的启发，EL 也设计成可以轻松地编写免脚本的 JSP 页面。也就是说，页面不使用任何 JSP 声明、表达式或者 scriptlet。第 11 章会进一步介绍为何无脚本的的 JSP 页面是一个最佳实践。

本章介绍如何使用 EL 表达式在 JSP 页面中显示数据和对象属性。它涵盖了最新的 EL 3.0 版本技术。

本章中的所有示例都可以在本书附带的 zip 文件中的 el-demo 项目中找到。

8.1 表达式语言简史

JSP 2.0 最初是将 EL 应用在 JSP 标准标签库（JSTL）1.0 规范中。JSP 1.2 程序员将标准库导入到他们的应用程序中，就可以使用 EL。JSP 2.0 及其更高版本的用户即使没有 JSTL，也能使用 EL，但在许多应用程序中，还是需要 JSTL 的，因为它里面还包含了与 EL 无关的其他标签。

JSP 2.1 和 JSP 2.2 中的 EL 要将 JSP 2.0 中的 EL 与 JSF（JavaServer Faces）中定义的 EL 统一起来。JSF 是在 Java 中快速构建 Web 应用程序的框架，并且是构建在 JSP 1.2 之上。由于 JSP 1.2 中缺乏整合式的表达式语言，并且 JSP 2.0 EL 也无法满足 JSF 的所有需求，因此为 JSF 1.0 开发出了一款 EL 的变体。后来这两种语言变体合二为一。

2013 年 5 月发布了 EL 3.0 版本（JSR 341），EL 不再是 JSP 或任何其他技术的一部分，而是一个独立的规范。EL 3.0 添加了对 lambda 表达式的支持，并允许集合操作。其 lambda 支持不需要 Java SE 8，Java SE 7 即可。

8.2 表达式语言的语法

EL 表达式以 ${ 开头，并以 } 结束。EL 表达式的结构如下：

```
${expression}
#{expression}
```

例如，表达式 x+y，可以写成：

```
${x+y}
```

或

```
#{x+y}
```

$ {exp}和＃{exp}结构都由 EL 引擎以相同的方式进行计算。然而，当 EL 未被用作独立引擎而是使用诸如 JSF 或 JSP 的底层技术时，该技术可以不同地解释构造。例如，在 JSF 中，${exp}结构用于立即计算，＃{expr}结构用于延迟计算（即表达式直到系统需要它的值时，才进行计算）。另一方面，立即计算的表达式，会在 JSP 页面编译时同时编译，并在执行 JSP 页面时被执行。在 JSP 2.1 和更高版本中，＃{exp}表达式只能在接受延迟表达式的标签属性中使用。

两个表达式可以连接在一起。对于一系列的表达式，它们的取值将是从左到右进行，计算结果的类型为 String，并且连接在一起。假如 *a*+*b* 等于 8，*c*+*d* 等于 10，那么这两个表达式的计算结果将是 810：

```
${a+b}${c+d}
```

表达式${a+b}and${c+d}的取值结果则是 8and10。

如果在定制标签的属性值中使用 EL 表达式，那么该表达式的取值结果字符串将会强制变成该属性需要的类型：

```
<my:tag someAttribute="${expression}"/>
```

像${这样的字符顺序就表示是一个 EL 表达式的开头。如果需要的只是文本${，则需要在它前面加一个转义符，如\${。

8.2.1 关键字

以下是关键字，它们不能用作标识符：

```
and  eq  gt  true  instanceof
```

```
or   ne  le   false  empty
not  lt   ge   null  div  mod
```

8.2.2 [ ]和.运算符

EL 表达式可以返回任意类型的值。如果 EL 表达式的结果是一个带有属性的对象，则可以利用[ ]或者.运算符来访问该属性。[ ]和.运算符类似；[ ]是比较规范的形式，.运算符则比较快捷。

为了访问对象的属性，可以使用以下任意一种形式：

```
${object["propertyName"]}
${object.propertyName}
```

但是，如果 propertyName 不是有效的 Java 变量名，只能使用[ ]运算符。例如，下面这两个 EL 表达式就可以用来访问隐式对象标题中的 HTTP 标题 host：

```
${header["host"]}
${header.host}
```

但是，要想访问 accept-language 标题，只能使用[ ]运算符，因为 accept-language 不是一个合法的 Java 变量名。如果用. 运算符访问它，将会导致异常。

如果对象的属性碰巧返回带有属性的另一个对象，既可以用[ ]，也可以用. 运算符来访问第二个对象的属性。例如，隐式对象 pageContext 是表示当前 JSP 的 PageContext 对象。它有 request 属性，表示 HttpServletRequest。HttpServletRequest 带有 servletPath 属性。那么，下列几个表达式的结果相同，均能得出 pageContext 中 HttpServletRequest 的 servletPath 属性值：

```
${pageContext["request"]["servletPath"]}
${pageContext.request["servletPath"]}
${pageContext.request.servletPath}
${pageContext["request"].servletPath}
```

要访问 HttpSession，可以使用以下语法：

```
${pageContext.session}
```

例如，以下表达式会得出 session 标识符。

```
${pageContext.session.id}
```

8.2.3 取值规则

EL 表达式的取值是从左到右进行的。对于 expr-a[expr-b]形式的表达式，其 EL 表达式的取值方法如下：

（1）先计算 expr-a 得到 value-a。

（2）如果 value-a 为 null，则返回 null。

（3）然后计算 expr-b 得到 value-b。

（4）如果 value-b 为 null，则返回 null。

（5）如果 value-a 为 java.util.Map，则会查看 value-b 是否为 Map 中的一个 key。若是，则返回 value-a.get(value-b)，若不是，则返回 null。

（6）如果 value-a 为 java.util.List，或者假如它是一个 array，则要进行以下处理：

a．强制 value-b 为 int，如果强制失败，则抛出异常。

b．如果 value-a.get(value-b)抛出 IndexOutOfBoundsException，或者假如 Array.get (value-a, value-b)抛出 ArrayIndexOutOfBoundsException，则返回 null。

c．否则，若 value-a 是个 List，则返回 value-a.get(value-b)；若 value-a 是个 array，则返回 Array.get(value-a, value-b)。

（7）如果 value-a 不是一个 Map、List 或者 array，那么，value-a 必须是一个 JavaBean。在这种情况下，必须强制 value-b 为 String。如果 value-b 是 value-a 的一个可读属性，则要调用该属性的 getter 方法，从中返回值。如果 getter 方法抛出异常，该表达式就是无效的，否则，该表达式有效。

8.3　访问 JavaBean

利用. 或[]运算符，都可以访问 bean 的属性，其结构如下：

```
${beanName["propertyName"]}
${beanName.propertyName}
```

例如，访问 myBean 的 secret 属性，可以使用以下表达式：

```
${myBean.secret}
```

如果该属性是一个带属性的对象，那么同样也可以利用.或[]运算符来访问第二个对象的该属性。假如该属性是一个 Map、List 或者 array，则可以利用 8.2 节介绍的访问 Map 值或 List 成员或 array 元素的同样规则。

8.4 EL 隐式对象

在 JSP 页面中，可以利用 JSP 脚本来访问 JSP 隐式对象。但是，在免脚本的 JSP 页面中，则不可能访问这些隐式对象。EL 允许通过提供一组它自己的隐式对象来访问不同的对象。EL 隐式对象见表 8.1。

表 8.1 EL 隐式对象

对象	描述
pageContext	这是当前 JSP 的 javax.servlet.jsp.PageContext
initParam	这是一个包含所有环境初始化参数并用参数名作为 key 的 Map
param	这是一个包含所有请求参数并用参数名作为 key 的 Map。每个 key 的值就是指定名称的第一个参数值。因此，如果两个请求参数同名，则只有第一个能够利用 param 获取值。要想访问同名参数的所有参数值，可用 params 代替
paramValues	这是一个包含所有请求参数并用参数名作为 key 的 Map。每个 key 的值就是一个字符串数组，其中包含了指定参数名称的所有参数值。就算该参数只有一个值，它也仍然会返回一个带有一个元素的数组
header	这是一个包含请求标题并用标题名作为 key 的 Map。每个 key 的值就是指定标题名称的第一个标题。换句话说，如果一个标题的值不止一个，则只返回第一个值。要想获得多个值的标题，得用 headerValues 对象代替
headerValues	这是一个包含请求标题并用标题名作为 key 的 Map。每个 key 的值就是一个字符串数组，其中包含了指定标题名称的所有参数值。就算该标题只有一个值，它也仍然会返回一个带有一个元素的数组
cookie	这是一个包含了当前请求对象中所有 Cookie 对象的 Map。Cookie 名称就是 key 名称，并且每个 key 都映射到一个 Cookie 对象
applicationScope	这是一个包含了 ServletContext 对象中所有属性的 Map，并用属性名称作为 key
sessionScope	这是一个包含了 HttpSession 对象中所有属性的 Map，并用属性名称作为 key
requestScope	这是一个 Map，其中包含了当前 HttpServletRequest 对象中的所有属性，并用属性名称作为 key
pageScope	这是一个 Map，其中包含了全页面范围内的所有属性。属性名称就是 Map 的 key

下面逐个介绍这些对象。

8.4.1 pageContext

pageContext 对象表示当前 JSP 页面的 javax.servlet.jsp.PageContext。它包含了所有其他的 JSP 隐式对象，见表 8.2。

表 8.2　JSP 隐式对象

对象	EL 中的类型
request	javax.servlet.http.HttpServletRequest
response	javax.servlet.http.HttpServletResponse
out	javax.servlet.jsp.JspWriter
session	javax.servlet.http.HttpSession
application	javax.servlet.ServletContext
config	javax.servlet.ServletConfig
PageContext	javax.servlet.jsp.PageContext
page	javax.servlet.jsp.HttpJspPage
exception	java.lang.Throwable

例如，可以利用以下任意一个表达式来获取当前的 ServletRequest：

```
${pageContext.request}
${pageContext["request"]
```

并且，还可以利用以下任意一个表达式来获取请求方法：

```
${pageContext["request"]["method"]}
${pageContext["request"].method}
${pageContext.request["method"]}
${pageContext.request.method}
```

表 8.3 列出了 pageContext.request 中一些有用的属性。

表 8.3　pageContext.request 中一些有用的属性

属性	说明
characterEncoding	请求的字符编码
contentType	请求的 MIME 类型
locale	浏览器首先 locale
locales	所有 locale
protocol	HTTP 协议，例如：HTTP/1.1
remoteAddr	客户端 IP 地址
remoteHost	客户端 IP 地址或主机名
scheme	请求发送方案，HTTP 或 HTTPS
serverName	服务器主机名
serverPort	服务器端口
secure	请求是否通过安全链接传输

对请求参数的访问比对其他隐式对象更加频繁；因此，这里提供了 param 和 paramValues 两个隐式对象。

8.4.2 initParam

隐式对象 initParam 用于获取上下文参数的值。例如，为了获取名为 password 的上下文参数值，可以使用以下表达式：

```
${initParam.password}
```

或者

```
${initParam["password"]
```

8.4.3 param

隐式对象 param 用于获取请求参数值。这个对象表示一个包含所有请求参数的 Map。例如，要获取 userName 参数，可以使用以下任意一种表达式：

```
${param.userName}
${param["userName"]}
```

8.4.4 paramValues

利用隐式对象 paramValues 可以获取一个请求参数的多个值。这个对象表示一个包含所有请求参数，并以参数名称作为 key 的 Map。每个 key 的值是一个字符串数组，其中包含了指定参数名称的所有值。即使该参数只有一个值，它也仍然返回一个带有一个元素的数组。例如，为了获得 selectedOptions 参数的第一个值和第二个值，可以使用以下表达式：

```
${paramValues.selectedOptions[0]}
${paramValues.selectedOptions[1]}
```

8.4.5 header

隐式对象 header 表示一个包含所有请求标题的 Map。为了获取 header 值，要利用 header 名称作为 key。例如，为了获取 accept-language 这个 header 值，可以使用以下表达式：

```
${header["accept-language"]}
```

如果 header 名称是一个有效的 Java 变量名，如 connection，那么也可以使用．运算符：

```
${header.connection}
```

8.4.6 headerValues

隐式对象 headerValues 表示一个包含所有请求 head 并以 header 名称作为 key 的 Map。但是，与 head 不同的是，隐式对象 headerValues 返回的 Map 返回的是一个字符串数组。例如，为了获取标题 accept-language 的第一个值，要使用以下表达式：

```
${headerValues["accept-language"][0]}
```

8.4.7 cookie

隐式对象 cookie 可以用来获取一个 cookie。这个对象表示当前 HttpServletRequest 中所有 cookie 的值。例如，为了获取名为 jsessionid 的 cookie 值，要使用以下表达式：

```
${cookie.jsessionid.value}
```

为了获取 jsessionid cookie 的路径值，要使用以下表达式：

```
${cookie.jsessionid.path}
```

8.4.8 applicationScope、sessionScope、requestScope 和 pageScope

隐式对象 applicationScope 用于获取应用程序范围级变量的值。假如有一个应用程序范围级变量 myVar，就可以利用以下表达式来获取这个属性：

```
${applicationScope.myVar}
```

注意，在 servlet/JSP 编程中，有界对象是指在以下对象中作为属性的对象：PageContext、ServletRequest、HttpSession 或者 ServletContext。隐式对象 sessionScope、requestScope 和 pageScope 与 applicationScope 相似。但是，其范围分别为 session、request 和 page。

有界对象也可以通过没有范围的 EL 表达式获取。在这种情况下，JSP 容器将返回 PageContext、ServletRequest、HttpSession 或者 ServletContext 中第一个同名的对象。执行顺序是从最小范围（PageContext）到最大范围（ServletContext）。例如，以下表达式将返回 today 引用的任意范围的对象。

```
${today}
```

8.5 使用其他 EL 运算符

除了. 和 [] 运算符外，EL 还提供了其他运算符：算术运算符、关系运算符、逻辑运算符、条件运算符，以及 empty 运算符。使用这些运算符时，可以进行不同的运算。但是，由于 EL 的目的是方便免脚本 JSP 页面的编程，因此，除了关系运算符外，这些 EL 运算符的用处都很有限。

8.5.1 算术运算符

算术运算符有 5 种。

- 加法（+）。
- 减法（−）。
- 乘法（*）。
- 除法（/和 div）。
- 取余/取模（%和 mod）。

除法和取余运算符都有两种形式，与 XPath 和 ECMAScript 是一致的。

注意，EL 表达式的计算按优先级从高到低、从左到右进行。下列运算符是按优先级递减顺序排列的：

- */div%mod
- +-

这表示*、/、div、%以及 mod 运算符的优先级相同，+与−的优先级相同，但第二组运算符的优先级小于第一组运算符。因此，表达式

```
${1+2*3}
```

的运算结果是 7，而不是 9。

8.5.2 关系运算符

下面是关系运算符列表：

- 等于（==和 eq）。
- 不等于（!=和 ne）。
- 大于（>和 gt）。
- 大于或等于（>=和 ge）。

- 小于（<和 lt）。
- 小于或等于（<=和 le）。

例如，表达式${3==4}返回 False，${“b”<“d”}则返回 True。

8.5.3 逻辑运算符

下面是逻辑运算符列表：

- 和（&&和 and）。
- 或（|| 和 or）。
- 非（！ 和 not）。

8.5.4 条件运算符

EL 条件运算符的语法如下：

```
${statement? A:B}
```

如果 statement 的计算结果为 True，那么该表达式的输出结果就是 A，否则为 B。

例如，利用下列 EL 表达式可以测试 HttpSession 中是否包含名为 loggedIn 的属性。如果找到这个属性，就显示“You have logged in（您已经登录）”。否则显示“You have not logged in（您尚未登录）”。

```
${(sessionScope.loggedIn==null)? "You have not logged in" :
        "You have logged in"}
```

8.5.5 empty 运算符

empty 运算符用来检查某一个值是否为 null 或者 empty。下面是一个 empty 运算符的使用范例：

```
${empty X}
```

如果 X 为 null，或者说 X 是一个长度为 0 的字符串，那么该表达式将返回 True。如果 X 是一个空 Map、空数组或者空集合，它也将返回 True。否则，将返回 False。

8.5.6 字符串连接运算符

+=运算符用于连接字符串。 例如，以下表达式打印 a + b 的值。

```
$ {a + = b}
```

8.5.7 分号操作符

; 操作符用于分隔两个表达式。 有关示例，请参阅本章后面的 8.10 节。

8.6 引用静态属性和静态方法

您可以引用在任何 Java 类中定义的静态字段和方法。 但是，在可以在 JSP 页面中引用静态字段或方法之前，必须使用 page 伪指令导入类或类包。 java.lang 包是一个例外，因为它是自动导入的。

例如，以下 page 指令导入 java.time 包。

```
<%@page import="java.time.*"%>
```

或者，你可以导入单个类，例如

```
<%@page import="java.time.LocalDate"%>
```

限制可以引用 LocalDate 类的静态方法：now 方法，如下：

```
Today is ${LocalDate.now()}
```

在本章后面的章节中，我们将学习如何格式化日期。

如下是引用类的静态成员和静态字段的另一个例子。

```
<p>
    &radic; <span style = “text-decoration: overline;” >  36  </ span>
    = $ {Math.sqrt ( 36 ) }
</p>
<p>
    &pi; = ${Math.PI}
</p>
```

Math 是 java.lang 包下类，无需额外的导入。图 8.1 展示了以上代码的 JSP 页面效果。

此外，还有一种导入包的方法，是在 ServletContextlistener 中以编程方式导入。清单 8.1 显示了一个监听器，它导入两个包 java.time 和 java.util。

$\sqrt{36} = 6.0$

π = 3.141592653589793

图 8.1　引用类静态成员

清单 8.1　编程导入类型

```
package listener;
import javax.el.ELContextEvent;
import javax.servlet.ServletContextEvent;
import javax.servlet.ServletContextListener;
import javax.servlet.annotation.WebListener;
import javax.servlet.jsp.JspFactory;

@WebListener
public class ELImportListener implements ServletContextListener {

    @Override
    public void contextInitialized(ServletContextEvent event) {
        JspFactory.getDefaultFactory().getJspApplicationContext(
                event.getServletContext()).addELContextListener(
                        (ELContextEvent e) -> {
                                e.getELContext().getImportHandler().
                                importPackage("java.time");
                                e.getELContext().getImportHandler().
                                importPackage("java.util");
                        });
    }

    @Override
    public void contextDestroyed(ServletContextEvent event) {
    }
}
```

8.7　创建 Set、List 和 Map

可以动态的创建 Set、List 和 Map。创建一个 Set 的语法如下：

```
{ comma-delimited-elements }
```

例如，如下表达式创建一个 5 个数字 Set：

```
${{1, 2, 3, 4, 5}}
```

创建一个 List 的语法如下

```
[ comma-delimited-elements ]
```

例如，如下表达式创建了一组花名的 List：

```
${["Aster", "Carnation", "Rose"]}
```

最后，创建一个 Map 的语法为：

```
{ comma-delimited-key-value-entries}
```

如下为一组国家及其首都的 Maps：

```
${{"Canada": "Ottawa", "China": "Beijing", "France": "Paris"}}
```

8.8 访问列表元素和 Map 条目

可以通过索引来访问 List，如下表达返回 Cities 的第一个元素

```
${cities[0]}
```

可以通过如下方式访问 Map

```
${map[key]}
```

例如，下面的表达式返回“Ottawa”：

```
${{"Canada": "Ottawa", "China": "Beijing"} ["Canada"]}
```

8.9 操作集合

EL 3.0 带来了很多新特性。其中一个主要的贡献是操纵集合的能力。你可以通过调用流方法将集合转换为流来使用此功能。

下面展示如何将列表转换为流，假设 myList 是一个 java.util.List：

```
${myList.stream()}
```

大部分流的操作会返回另一个流，因而可以形成链式操作。

```
$ {MyList.stream().operation-1().operation-2().toList()}
```

在链式操作的末尾，通常调用 toList 方法，以便打印或格式化结果。

以下小节介绍了你可以对流执行的一些操作

8.9.1 toList

toList 方法返回一个 List，它包含与当前流相同的成员。调用此方法的主要目的是轻松地打印或操作流元素。下面是一个将列表转换为流并返回列表的示例：

```
$ {[100, 200, 300].stream().toList()}
```

当然这个例子没有什么用。稍后在接下来的小节中，你将看到更多的例子。

8.9.2　toArray

与 toList 类似，但返回一个 Java 数组。同样，在数组中呈现元素通常是有用的，因为许多 Java 方法将数组作为参数。这里是一个 toArray 的例子：

```
${["One", "Two", "Three"].stream().toArray()}
```

与 toList 不同，toArray 不打印元素。因此，toList 更经常使用。

8.9.3　limit

limit 方法限制流中元素的数量。

名为 cities 的 List 包含 7 个城市：

```
[Paris, Strasbourg, London, New York, Beijing, Amsterdam, San Francisco]
```

下面的代码将元素的数量限制为 3。

```
${cities.stream().limit(3).toList()}
```

执行时，表达式将返回此列表：

```
[Paris, Strasbourg, London]
```

如果传递给 limit 方法的参数大于元素的数量，则返回所有元素。

8.9.4　sort

此方法对流中的元素进行排序。例如，这个表达式

```
${cities.stream().sorted().toList()}
```

返回如下排序后的列表。

```
[Amsterdam, Beijing, London, New York, Paris, San Francisco, Strasbourg]
```

8.9.5　average

此方法返回流中所有元素的平均值。其返回值是一个 Optional 对象，它可能为 null。需要调用 get()获取实际值。

此表达式返回 4.0：

```
${[1,3,5,7].stream().average().get()}
```

8.9.6 sum

此方法计算流中所有元素的总和。例如，此表达式返回 16。

```
${[1,3,5,7].stream().sum()}
```

8.9.7 count

此方法返回流中元素的数量。例如，此表达式返回 7。

```
${cities.stream().count()}
```

8.9.8 min

此方法返回流的元素中的最小值。同 average 方法一样，其返回值是一个 Optional 对象，因此你需要调用 get 方法来获取实际值。例如，此表达式返回 1。

```
${[1,3,100,1000].stream().min().get()}
```

8.9.9 max

此方法返回流的元素中的最大值。同 average 方法一样，其返回值是一个 Optional 对象，因此你需要调用 get 方法来获取实际值。例如，此表达式返回 1000。

```
${[1,3,100,1000].stream().max().get()}
```

8.9.10 map

此方法将流中的每个元素映射到另一个流中的另一个元素，并返回该流。此方法接受一个 lambda 表达式。

例如，此映射方法使用 lambda 表达式 x - > 2 * x，这实际上将每个元素乘 2，并将它们返回到新的流中。

```
${[1,3,5].stream().map(x - > 2 * x).toList()}
```

返回列表如下：

```
[2,6,10]
```

下面是另一个示例，它将字符映射为大写。

```
${cities.stream().map(x - > x.toUpperCase()).toList()}
```

它返回以下列表。

```
[PARIS，STRASBOURG，LONDON，NEW YORK，BEIJING，AMSTERDAM，SAN FRANCISCO]
```

8.9.11　filter

此方法根据 lambda 表达式过滤流中的所有元素，并返回包含结果的新流。

例如，以下表达式测试城市是否以“S”开头，并返回所有的结果。

```
$ {cities.stream().filter(x - > x.startsWith("S")).toList()}
```

它产生的列表如下所示：

```
[Strasbourg, San Francisco]
```

8.9.12　forEach

此方法对流中的所有元素执行操作。它返回 void。 例如，此表达式将城市中的所有元素打印到控制台。

```
$ {cities.stream().forEach(x - > System.out.println(x))}
```

8.10　格式化集合

由于 EL 定义了如何写表达式而不是函数，因此无法直接打印或格式化集合，毕竟，打印和格式化不是 EL 负责的领域。然而，打印和格式化是两个不能忽视的重要任务。

如果你是 EL 3.0 的新手，但熟悉 JSP，那么格式化集合最简单的方法就是使用第 6 章中讨论的 JSTL。然而，具有强大的功能的 EL 3.0 应足以解决这些问题，并允许我们完全抛弃 JSTL。

例如，你可以尝试使用 forEach，类似于 JSTL 的 forEach。以下代码可以在 Tomcat 8 上运行。

```
$ {cities.stream().forEach(x - > pageContext.out.println(x))}
```

遗憾的是，这在 GlassFish 4 中不起作用，所以 forEach 不能通用。

我最终想出的两个解决方案不像 forEach 那样优雅，但在所有主要的 servlet 容器上都可用。第一个解决方案适用于 Java SE 7。第二个解决方案可能比第一个更优雅，但只适用于 Java SE 8。

8.10.1　使用 HTML 注释

List 的字符串表示形式如下所示：

```
[element-1, element-2, ...]
```

现在，如果我想在 HTML 中呈现列表元素，需要这样写。

```
<ul>
    <li> element-1 </ li>
    <li> element-2 </ li>
    ...
</ul>
```

现在，你可能已经注意到每个元素必须转向<li> element-n </ li>。我怎么做？如果你一直密切关注，你可能仍然记得 map 方法可以用来转换每个元素。所以，我会有这样的内容：

```
$ {myList.stream().map(x - >"<li>"+ = x + ="</ li>").toList()}
```

它给了我这样的 List：

```
[<li> element-1 </ li>, <li> element-2 </ li>, ...]
```

足够接近，但仍然需要删除括号和逗号。遗憾的是，你不能控制列表的字符串表示。好在你可以使用 HTML 注释。

所以，这里是一个例子：

```
<ul>
<!-${cities.stream().map(x - >"--> <li>"+ = x + ="</ li> <!--").toList()}-->
</ul>
```

结果如下所示：

```
<ul>
<!--[--><li>Paris</li><!--, --><li>Strasbourg</li><!--, -->
<li>London</li><!--, --><li>New York</li><!--, -->
<li>Beijing</li><!--, --><li>Amsterdam</li><!--, -->
<li>San Francisco</li><!--]-->
</ul>
```

这有效地注释掉了括号和逗号。虽然结果看起来有点凌乱，但它是有效的 HTML，更重要的是，它能工作！

图 8.2 显示了页面显示结果。

- Paris
- Strasbourg
- London
- New York
- Beijing
- Amsterdam
- San Francisco

图 8.2 使用 HTML 注释来格式化集合

这里有另一个例子，这个例子用表格格式化显示地址信息。

```
<table>
    <tr><th>Street</th><th>City</th></tr>
    <!--${addresses.stream().map(a->"-->
<tr><td>"+=a.streetName+="</td><td>"+=a.city+="</td></tr><!--").toList()}
```

```
-->
</table>
```

8.10.2　使用 String.join()

这第二个解决方案之所以有效，因为 EL 3.0 允许你引用静态方法。在 Java 8 中，String 类新增了一些静态方法，其中一个方法是 join，这正是我寻找的解决方案。有两个重载 join 方法，但这里要用到的一个方法如下所示。

```
public static String join(CharSequence delimiter,
        Iterable <? extends CharSequence> elements)
```

此方法返回用指定分隔符连接在一起的 CharSequence 元素组成的字符串。而 java.util. Collection 接口正好扩展了 Iterable。因此，你可以将 Collection 传递给 join 方法。

例如，下面是如何将列表格式化为 HTML 有序列表：

```
${"<ol> <li>"+ = String.join("</ li> <li>", 城市) + ="< li> </ ol>"}
```

此表达式适用于至少有一个元素的集合。如果你可能要处理一个空集合，这里是一个更好的表达式：

```
${empty cities? "" : "<ol><li>"
        += String.join("</li><li>", cities. stream().sorted().toList())
        += "</li></ol>"}
```

8.11　格式化数字

要格式化数字，你可以再次利用 EL 3.0 允许引用静态方法的能力。String 类的 format 静态方法可以用来格式化数字。

例如，以下表达式返回带有两个小数点位的数字。

```
${String.format("%-10.2f%n", 125.178)}
```

更多格式化规则可以查阅 java.text.DecimalFormat 的 javadoc 文档。

8.12　格式化日期

可以通过 String.format()来格式化一个 date 或 time。例如：

```
${d = LocalDate.now().plusDays(2); String.format("%tB %te, %tY%n", d, d, d)}
```

首先计算 LocalDate.now().plusDays(2)，并将结果复制给变量 d，然后用 String.format()方法来格式化 LocalDate，引用了 3 次变量 d。

更多的格式化规则见：

https://docs.oracle.com/javase/tutorial/java/data/numberformat.html

8.13 如何在 JSP 2.0 及其更高版本中配置 EL

有了 EL、JavaBeans 和定制标签，就可以编写免脚本的 JSP 页面了。JSP 2.0 及其更高的版本中还提供了一个开关，可以使所有的 JSP 页面都禁用脚本。现在，软件架构师可以强制编写免脚本的 JSP 页面了。

另一方面，在有些情况下，可能还会需要在应用程序中取消 EL。例如，正在使用与 JSP 2.0 兼容的容器，却尚未准备升级到 JSP 2.0，那么就需要这么做。在这种情况下，可以关闭 EL 表达式的计算。

8.13.1 实现免脚本的 JSP 页面

为了关闭 JSP 页面中的脚本元素，要使用 jsp-property-group 元素以及 url-pattern 和 scripting- invalid 两个子元素。url-pattern 元素定义禁用脚本要应用的 URL 样式。下面展示如何将一个应用程序中所有 JSP 页面的脚本都关闭：

```
<jsp-config>
    <jsp-property-group>
        <url-pattern>*.jsp</url-pattern>
        <scripting-invalid>true</scripting-invalid>
    </jsp-property-group>
</jsp-config>
```

注意：

在部署描述符中只能有一个 jsp-config 元素。如果已经为禁用 EL 而定义了一个 jsp-property-group，就必须在同一个 jsp-config 元素下，为禁用脚本而编写 jsp- property-group。

8.13.2 禁用 EL 计算

在某些情况下，比如，当需要在 JSP 2.0 及其更高版本的容器中部署 JSP 1.2 应用程序时，可能就需要禁用 JSP 页面中的 EL 计算了。此时，出现的 EL 结构，就不会作为 EL 表达式进行计算。目前有两种方式可以禁用 JSP 中的 EL 计算。

第一种，可以将 page 指令的 isELIgnored 属性设为 True，像这样：

```
<%@ page isELIgnored="true" %>
```

isELIgnored 属性的默认值为 False。如果想在一个或者几个 JSP 页面中关闭 EL 表达式计算，建议使用 isELIgnored 属性。

第二种，可以在部署描述符中使用 jsp-property-group 元素。jsp-property-group 元素是 jsp-config 元素的子元素。利用 jsp-property-group 可以将某些设置应用到应用程序中的一组 JSP 页面中。

为了利用 jsp-property-group 元素禁用 EL 计算，还必须有 url-pattern 和 el-ignored 两个子元素。url-pattern 元素用于定义 EL 禁用要应用的 URL 样式。el-ignored 元素必须设为 True。

下面举一个例子，展示如何在名为 noEl.jsp 的 JSP 页面中禁用 EL 计算。

```
<jsp-config>
    <jsp-property-group>
        <url-pattern>/noEl.jsp</url-pattern>
        <el-ignored>true</el-ignored>
    </jsp-property-group>
</jsp-config>
```

也可以像下面这样，通过给 url-pattern 元素赋值*.jsp，来禁用一个应用程序中的所有 JSP 页面的 EL 计算：

```
<jsp-config>
    <jsp-property-group>
        <url-pattern>*.jsp</url-pattern>
        <el-ignored>true</el-ignored>
    </jsp-property-group>
</jsp-config>
```

无论是将其 page 指令的 isELIgnored 属性设为 True，还是其 URL 与 el-ignored 为 True 的 jsp-property-group 的 URL 模式相匹配，都将禁用 JSP 页面中的 EL 计算。假如将一个 JSP 页面中 page 指令的 isELIgnored 属性设为 False，但其 URL 与在部署描述符中禁用了 EL 计算的 JSP 页面的模式匹配，那么该页面的 EL 计算也将被禁用。

此外，如果使用的是与 Servlet 2.3 及其更低版本兼容的部署描述符，那么 EL 计算已经默认关闭，即便使用的是 JSP 2.0 及其更高版本的容器，也一样。

8.14　小结

EL 是 JSP 2.0 及其更高版本中最重要的特性之一。它有助于编写更简短、更高效的 JSP 页面，还能帮助编写免脚本的页面。本章介绍了如何利用 EL 来访问 JavaBeans 和隐式对象，还介绍了如何使用 EL 运算符。本章的最后一个小节介绍了如何在与 JSP 2.0 及其更高版本相关的容器中使用与 EL 相关的应用程序设置。

第 9 章 JSTL

JSP 标准标签库（JavaServer Pages Standard Tag Library，JSTL）是一个定制标签库的集合，用来解决像遍历 Map 或集合、条件测试、XML 处理，甚至数据库访问和数据操作等常见的问题。

本章要介绍的是 JSTL 中最重要的标签，尤其是访问有界对象、遍历集合，以及格式化数字和日期的那些标签。如果有兴趣进一步了解，可以在 JSTL 规范文档中找到所有 JSTL 标签的完整版说明。

注意

随着 Expression Language 3.0 的发布，所有 JSTL 核心标记都可以用 EL 表达式替代。然而，旧项目仍然包含 JSTL，因此掌握 JSTL 仍然会有帮助。

9.1 下载 JSTL

JSTL 目前的最新版本是 1.2，这是由 JSR-52 专家组在 JCP（www.jcp.org）上定义的，在 java.net 网站可以下载：

```
http://jstl.java.net
```

其中，JSTL API 和 JSTL 实现这两个软件是必需下载的。JSTL API 中包含 javax.servlet.jsp.jstl 包，里面包含了 JSTL 规范中定义的类型。JSTL 实现中包含了实现类。这两个 jar 文件都必须复制到应用 JSTL 的每个应用程序的 WEB-INF/lib 目录下。

9.2 JSTL 库

JSTL 是标准标签库，但它是通过多个标签库来暴露其行为的。JSTL 1.2 中的标签可以分成 5 类区域，见表 9.1。

表 9.1 JSTL 标签库

区域	子函数	URI	前缀
核心	变量支持	http://java.sun.com/jsp/jstl/core	c
	流控制		
	URL 管理		
	其他		
XML	核心	http://java.sun.com/jsp/jstl/xml	x
	流控制		
	转换		
国际化	语言区域	http://java.sun.com/jsp/jstl/fmt	fmt
	消息格式化		
	数字和日期格式化		
数据库	SQL	http://java.sun.com/jsp/jstl/sql	sql
函数	集合长度	http://java.sun.com/jsp/jstl/functions	fn
	字符串操作		

在 JSP 页面中使用 JSTL 库，必须通过以下格式使用 taglib 指令：

```
<%@ taglib uri="uri" prefix="prefix" %>
```

例如，要使用 Core 库，必须在 JSP 页面的开头处做以下声明：

```
<%@ taglib uri="http://java.sun.com/jsp/jstl/core" prefix="c" %>
```

这个前缀可以是任意的。但是，采用惯例能使团队的其他开发人员以及后续加入该项目的其他人员更容易熟悉这些代码。因此，建议使用预定的前缀。

注意：

本章中讨论的每一个标签都会在各自独立的小节中做详细的介绍，每一个标签的属性也都将列表说明。属性名称后面的星号（*）表示该属性是必需的。加号（+）表示该属性的 rtexprvalue 值为 True，这意味着该属性可以赋静态字符串或者动态值（Java 表达式、EL 表达式，或者通过<jsp:attribute>设置的值）。rtexprvalue 值为 False 时，表示该属性只能赋静态字符串的值。

注意：

JSTL 标签的 body content 可以为 empty、JSP 或者 tagdependent。

9.3　一般行为

下面介绍 Core 库中用来操作有界变量的 3 个一般行为：out、set 和 remove。

9.3.1　out 标签

在运算表达式时，out 标签是将结果输出到当前的 JspWriter。out 的语法有两种形式，即有 body content 和没有 body content。

```
<c:out value="value" [escapeXml="{true|false}"]
        [default="defaultValue"]/>
<c:out value="value" [escapeXml="{true|false}"]>
    default value
</c:out>
```

注意：

在标签的语法中，[]表示可选的属性。如果值带下划线，则表示为默认值。

out 的 body content 为 JSP。out 标签的属性见表 9.2。

表 9.2　out 标签的属性

属性	类型	描述
value*+	对象	要计算的表达式
escapeXml+	布尔	表示结果中的字符<、>、&、'和 "将被转换成相应的实体码，如<转换成 lt;等等。
default+	对象	默认值

例如，下列的 out 标签将输出有界变量 X 的值：

```
<c:out value="${x}"/>
```

默认情况下，out 会将特殊字符<、>、'、"和&分别编写成它们相应的字符实体码 <、>、'、"和&。

在 JSP 2.0 版本前，out 标签是用于输出有界对象值的最容易的方法。在 JSP 2.0 及其更高的版本中，除非需要对某个值进行 XML 转义，否则可以放心地使用 EL 表达式：

```
${x}
```

警告：如果包含一个或多个特殊字符的字符串没有进行 XML 转义，它的值就无法在浏览器中正常显示。此外，没有转义的特殊字符，会使网站易于遭受交叉网站的脚本攻击。例如，别人可以对它 post 一个能够自动执行的 JavaScript 函数/表达式。

out 中的 default 属性可以赋一个默认值，当赋予其 value 属性的 EL 表达式返回 null 时，就会显示默认值。default 属性可以赋动态值，如果这个动态值返回 null，out 就会显示一个空的字符串。

例如，在下面的 out 标签中，如果在 HttpSession 中没有找到 myVar 变量，就会显示应用程序范围的变量 myVar 值。如果没有找到，则输出一个空的字符串。

```
<c:out value="${sessionScope.myVar}"
       default="${applicationScope.myVar"/>
```

9.3.2 url 标签

url 标签是非常有用的。简而言之，url 标签执行以下任一操作：

- 如果当前上下文路径为“/”（即应用程序部署为默认上下文），则它将空字符串附加到指定的路径。
- 如果当前上下文路径不是“/”，它会将上下文路径添加到指定的路径。

本节将通过以下一个小的应用程序来解释 url 标签的重要性，其结构在图 9.1 中给出。

该应用程序由两个 JSP 页面 main.jsp 和 admin.jsp 组成。main.jsp 文件位于应用程序根目录中，admin.jsp 文件位于 admin 文件夹中。二者都需要显示在图像文件夹中的两个图像，image1.png 和 image2.png。请注意，图像的绝对路径如下。

```
http: //host/context/image/image1.png
http: //host/context/image/image2.png
```

因为两个图像从不同位置被引用多次，为了方便使用，用一个包含文件来引用它们。任何需要显示图像的 JSP 页面仅需要将该包含文件添加到文件中即可。include 文件夹表明了有 4 个 JSP 页面这么做了。

```
webapp
  admin
    admin.jsp
  image
    image1.png
    image2.png
  include
    inc1.jspf
    inc2.jspf
    inc3.jspf
    inc4.jspf
  WEB-INF
    classes
    lib
      javax.servlet.jsp.jstl-1.2.1.jar
      javax.servlet.jsp.jstl-api-1.2.1.jar
  main.jsp
```

图 9.1 demo 应用的目录结构

清单 9.1 展示了第一个包含文件 inc1.jspf 的内容。

清单 9.1 inc1.jsp 的内容

```
inc1.jsp
<img src="image/image1.png"/>
<img src="../image/image2.png"/
```

第一个包含文件包含路径是相对于当前页的路径。假设 main.jsp 页面的 URL 是 http://host/context/main.jsp，那么这两个图像的 URL 将被解析为以下 URL：

```
http://host/context/image/image1.png
http://host/context/../image/image2.png
```

不难想象，结果不太令人满意。第一个图像能正确显示，但第二个图像不能。

当通过 http://host/context/admin/admin.jsp 访问管理页面时，图像的 URL 解析如下：

```
http://host/context/admin/image/image1.png
http://host/context/admin/../image/image2.png
```

结果，第一个图像将不显示，但第二个图像将显示。

很明显，使用相对路径并不总能工作，因为可以从不同目录中的 JSP 页面调用包含文件。我们唯一的希望是使图像 URL 相对于应用程序本身。所以就有了第二个包含文件：inc2.jspf（见清单 9.2）。

清单 9.2：inc2.jspf 的内容

```
inc2.jsp
<img src="/image/image1.png"/>
<img src="/image/image2.png"/>
```

这看起来不错，应该能工作了，对吧？遗憾的是，它并不适用于所有情况。这是因为在开发应用程序时，部署上下文路径通常是未知的。根据应用程序是否部署为默认上下文，admin.jsp 页面可能具有以下 URL 之一：

```
http://localhost/jstl-demo/admin/admin.jsp
http://localhost/admin/admin.jsp
```

在这两种情况下，浏览器不知道上下文路径。事实上，在第一个 URL 的情况下，它认为应用程序被部署为默认上下文，jstl-demo 是一个目录。因此，它将解析到第一个图像的 URL 为 http://localhost/image/image1.png，这使得图像无法显示。

仅当应用程序已部署到默认上下文时，才能正确显示这两个图像。然而，这足以说明 inc2.jspf 不是正确的解决方案。

url 标签是我们的救星。先看下使用了 url 标签的 inc3.jspf 的内容，见清单 9.3。

清单 9.3：inc3.jspf 的内容

```
<%@ taglib uri ="http://java.sun.com/jsp/jstl/core"prefix ="c"%>
inc3.jsp
<img src ="<c: url value ="/image/image1.png"/>"/>
```

```
<img src ="<c: url value ="/image/image2.png"/>"/>
```

这解决了问题，因为 url 标签在服务器上执行，它知道上下文路径是什么。所以，它可以正确地解析图像的路径。

还可以使用 EL 表达式，如清单 9.4 中的 inc4.jspf 所示。

清单 9.4：inc4.jspf 的内容

```
inc4.jsp
<! - $ {cp = pageContext.request.contextPath} -->
<img src ="${cp =="/"? " ": cp}/image/image1.png"/>
<img src ="${cp =="/"? " ": cp}/image/image2.png"/>
```

可以使用以下表达式获取上下文路径：

```
${pageContext.request.contextPath}
```

相同的表达式将多次使用，所以我还创建了一个快捷方式 cp。

```
${cp = pageContext.request.contextPath}
```

然而，该值仍然会发送到浏览器，这就是为什么它被放在一个 HTML 注释中。然后，你只需要测试上下文路径是“/”（默认上下文）还是别的东西。

```
${cp =="/"?  "": cp}
```

图 9.2 显示了非默认上下文中的 admin.jsp 页面。

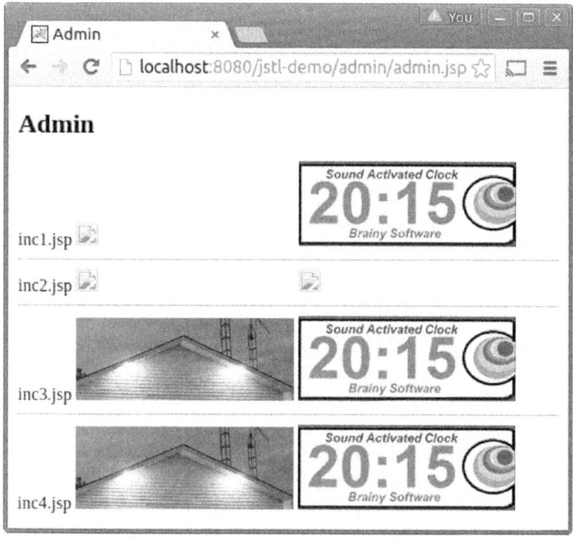

图 9.2　图像引用的四种不同方式

正如你可以看到的，只有 inc3.jspf 和 inc4.jspf（分别采用 url 标签和 EL）能够工作。

9.3.3　set 标签

利用 set 标签，可以完成以下工作：

（1）创建一个字符串和一个引用该字符串的有界变量。

（2）创建一个引用现存有界对象的有界变量。

（3）设置有界对象的属性。

如果用 set 创建有界变量，那么，在该标签出现后的整个 JSP 页面中都可以使用该变量。

set 标签的语法有 4 种形式。第一种形式用于创建一个有界变量，并用 value 属性在其中定义一个要创建的字符串或者现存有界对象。

```
<c:set value="value" var="varName"
        [scope="{page|request|session|application}"]/>
```

这里的 scope 属性指定了有界变量的范围。

例如，下面的 set 标签创建了字符串“The wisest fool”，并将它赋给新创建的页面范围变量 foo：

```
<c:set var="foo" value="The wisest fool"/>
```

下面的 set 标签则创建了一个名为 job 的有界变量，它引用 requestScope 对象 position。变量 job 的范围为 page。

```
<c:set var="job" value="${requestScope.position}" scope="page"/>
```

注意：

最后一个例子可能有点令人费解，因为它创建了一个引用请求范围对象的页面范围变量。如果清楚有界对象本身并非真的在 HttpServletRequest“里面”，就不难明白了。引用（名为 position）其实是指引用该对象。有了上一个例子中的 set 标签，再创建一个引用相同对象的有界变量（job）即可。

第二种形式与第一种形式相似，只是要创建的字符串或者要引用的有界对象是作为 body content 赋值的。

```
<c:set var="varName" [scope="{page|request|session|application}"]>
    body content
</c:set>
```

第二种形式允许在 body content 中有 JSP 代码。

第三种形式是设置有界对象的属性值。target 属性定义有界对象以及有界对象的 property 属性。对该属性的赋值是通过 value 属性进行的。

```
<c:set target="target" property="propertyName" value="value"/>
```

例如，下面的 set 标签是将字符串“Tokyo”赋予有界对象 address 的 city 属性。

```
<c:set target="${address}" property="city" value="Tokyo"/>
```

注意，必须在 target 属性中用一个 EL 表达式来引用这个有界对象。

第四种形式与第三种形式相似，只是赋值是作为 body content 完成的。

```
<c:set target="target" property="propertyName">
    body content
</c:set>
```

例如，下面的 set 标签是将字符串“Beijing”赋予有界对象 address 的 city 属性。

```
<c:set target="${address}" property="city">Beijing</c:set>
```

set 标签的属性见表 9.3。

表 9.3　set 标签的属性

属性	类型	描述
value+	对象	要创建的字符串，或者要引用的有界对象，或者新的属性值
var	字符串	要创建的有界变量
scope	字符串	新创建的有界变量的范围
target+	对象	其属性要被赋新值的有界对象；这必须是一个 JavaBeans 实例或者 java.util.Map 对象
property+	字符串	要被赋新值的属性名称

9.3.4　remove 标签

remove 标签用于删除有界变量，其语法如下：

```
<c:remove var="varName"
        [scope="{page|request|session|application}"]/>
```

注意，有界变量引用的对象不能删除。因此，如果另一个有界对象也引用了同一个对象，仍然可以通过另一个有界变量访问该对象。

remove 标签的属性见表 9.4。

表 9.4　remove 标签的属性

属性	类型	描述
var	字符串	要删除的有界变量的名称
scope	字符串	要删除的有界变量的范围

举个例子，下面的 remove 标签删除了页面范围的变量 job。

```
<c:remove var="job" scope="page"/>
```

9.4 条件行为

条件行为用于处理页面输出取决于特定输入值的情况，这在 Java 中是利用 if、 if …else 和 switch 声明解决的。

JSTL 中执行条件行为的有 4 个标签，即 if、choose、when 和 otherwise 标签。下面分别对其进行详细讲解。

9.4.1 if 标签

if 标签是对某一个条件进行测试，假如结果为 True，就处理它的 body content。测试结果保存在 Boolean 对象中，并创建有界变量来引用这个 Boolean 对象。利用 var 属性和 scope 属性分别定义有界变量的名称和范围。

if 的语法有两种形式。第一种形式没有 body content。

```
<c:if test="testCondition" var="varName"
        [scope="{page|request|session|application}"]/>
```

在这种情况下，var 定义的有界对象一般是由其他标签在同一个 JSP 的后续阶段进行测试。

第二种形式中使用了一个 body content。

```
<c:if test="testCondition [var="varName"]
        [scope="{page|request|session|application}"]>
    body content
</c:if>
```

body content 是 JSP，当测试条件的结果为 True 时，就会得到处理。if 标签的属性见表 9.5。

表 9.5 if 标签的属性

属性	类型	描述
test+	布尔	决定是否处理任何现有 body content 的测试条件
var	字符串	引用测试条件值的有界变量名称；var 的类型为 Boolean
scope	字符串	var 定义的有界变量的范围

例如，如果找到请求参数 user 且值为 ken，并且找到请求参数 password 且值为 blackcomb，以下 if 标签将显示“You logged in successfully（你已经成功登录）”：

```
<c:if test="${param.user=='ken' && param.password=='blackcomb'}">
    You logged in successfully.
</c:if>
```

为了模拟 else，下面使用了两个 if 标签，并使用了相反的条件。例如，如果 user 和 password 参数的值为 ken 和 blackcomb，以下代码片断将显示“You logged in successfully（你已经成功登录）”。否则，将显示“Login failed（登录失败）”。

```
<c:if test="${param.user=='ken' && param.password=='blackcomb'}">
    You logged in successfully.
</c:if>
<c:if test="${!(param.user=='ken' && param.password=='blackcomb')}">
    Login failed.
</c:if>
```

下面的 if 标签测试 user 和 password 参数值是否分别为 ken 和 blackcomb，并将结果保存在页面范围的变量 loggedIn 中。之后，利用一个 EL 表达式，如果 loggedIn 变量值为 True，则显示“You logged in successfully（你已经成功登录）”；如果 loggedIn 变量值为 False，则显示“Login failed（登录失败）”。

```
<c:if var="loggedIn"
        test="${param.user=='ken' && param.password=='blackcomb'}"/>
    ...
${(loggedIn)? "You logged in successfully" : "Login failed"}
```

9.4.2 choose、when 和 otherwise 标签

choose 和 when 标签的作用与 Java 中的关键字 switch 和 case 类似。也就是说，它们是用于为相互排斥的条件执行提供上下文的。choose 标签中必须嵌有一个或者多个 when 标签，并且每个 when 标签都表示一种可以计算和处理的情况。otherwise 标签则用于默认的条件块，假如没有任何一个 when 标签的测试条件结果为 True，otherwise 就会得到处理。假如是这种

情况，otherwise 就必须放在最后一个 when 之后。

choose 和 otherwise 标签没有属性。when 标签必须带有定义测试条件的 test 属性，用来决定是否应该处理 body content。

举个例子，以下代码是测试参数 status 的值。如果 status 的值为 full，将显示“You are a full member（你是正式会员）”。如果这个值为 student，则显示 “You are a student member（你是学生会员）”。如果 status 参数不存在，或者它的值既不是 full，也不是 student，那么这段代码将不显示任何内容。

```
<c:choose>
    <c:when test="${param.status=='full'}">
        You are a full member
    </c:when>
    <c:when test="${param.status=='student'}">
        You are a student member
    </c:when>
</c:choose>
```

下面的例子与前面的例子相似，但它是利用 otherwise 标签，如果 status 参数不存在，或者它的值不是 full 或 student，则将显示 “Please register（请注册）”。

```
<c:choose>
    <c:when test="${param.status=='full'}">
        You are a full member
    </c:when>
    <c:when test="${param.status=='student'}">
        You are a student member
    </c:when>
    <c:otherwise>
        Please register
    </c:otherwise>
</c:choose>
```

9.5 遍历行为

当需要无数次地遍历一个对象集合时，遍历行为就很有帮助。JSTL 提供了 forEach 和 forTokens 两个执行遍历行为的标签，这两个标签将在接下来的小节中介绍。

9.5.1 forEach 标签

forEach 标签会无数次地反复遍历 body content 或者对象集合。可以遍历的对象包括

java.util.Collection 和 java.util.Map 的所有实现，以及对象数组或者基本类型。也可以遍历 java.util.Iterator 和 java.util.Enumeration，但不应该在多个行为中使用 Iterator 或者 Enumeration，因为无法重置 Iterator 或者 Enumeration。

forEach 标签的语法有两种形式。第一种形式是固定次数地重复 body content。

```
<c:forEach [var="varName"] begin="begin" end="end" step="step">
    body content
</c:forEach>
```

第二种形式用于遍历对象集合。

```
<c:forEach items="collection" [var="varName"]
        [varStatus="varStatusName"] [begin="begin"] [end="end"]
        [step="step"]>
    body content
</c:forEach>
```

body content 是 JSP。forEach 标签的属性见表 9.6。

表 9.6 forEach 标签的属性

属性	类型	描述
var	字符串	引用遍历的当前项目的有界变量名称
items+	支持的任意类型	遍历的对象集合
varStatus	字符串	保存遍历状态的有界变量名称。类型值为 javax.servlet.jsp.jstl.core.LoopTagStatus
begin+	整数	如果指定 items，遍历将从指定索引处的项开始，例如，集合中第一项的索引为 0。如果没有指定 items，遍历将从设定的索引值开始。如果指定，begin 的值必须大于或者等于 0
end+	整数	如果指定 items，遍历将在（含）指定索引处的项结束。如果没有指定 items，遍历将在索引到达指定值时结束
step+	整数	遍历将只处理间隔指定 step 的项目，从第一项开始。在这种情况下，step 的值必须大于或者等于 1

例如，下列的 forEach 标签将显示“1，2，3，4，5”。

```
<c:forEach var="x" begin="1" end="5">
    <c:out value="${x}"/>,
</c:forEach>
```

下面的 forEach 标签将遍历有界变量 address 的 phones 属性：

```
<c:forEach var="phone" items="${address.phones}">
    ${phone}"<br/>
```

```
</c:forEach>
```

对于每一次遍历，forEach 标签都将创建一个有界变量，变量名称通过 var 属性定义。在本例中，有界变量命名为 phone。forEach 标签中的 EL 表达式用于显示 phone 的值。这个有界变量只存在于开始和结束的 forEach 标签之间，一到结束的 forEach 标签前，它就会被删除。

forEach 标签有一个类型为 javax.servlet.jsp.jstl.core.LoopTagStatus 的变量 varStatus。LoopTagStatus 接口带有 count 属性，它返回当前遍历的“次数”。第一次遍历时，status.count 值为 1；第二次遍历时，Status.count 值为 2，依此类推。通过测试 status.count%2 的余数，可以知道该标签正在处理的是偶数编号的元素，还是奇数编号的元素。

以 iterator 应用程序中的 BookServlet 类和 BookList.jsp 页面为例。如清单 9.5 所示，BookServer 类在其 doGet 方法中，创建了 3 个 Book 对象，并放入到一个 List 对象中，并将该 List 对象存入 ServletRequest 属性中，最后转到 books.jsp 页面，该页面会通过 forEach 标签遍历图书集合。Book 类如清单 9.6 所示。

清单 9.5　Books Servlet 类

```
package servlet;
import java.io.IOException;
import java.util.ArrayList;
import java.util.List;
import javax.servlet.RequestDispatcher;
import javax.servlet.ServletException;
import javax.servlet.annotation.WebServlet;
import javax.servlet.http.HttpServlet;
import javax.servlet.http.HttpServletRequest;
import javax.servlet.http.HttpServletResponse;
import app05a.model.Book ;

@WebServlet(urlPatterns = {"/books"})
public class BooksServlet extends HttpServlet {
    private static final int serialVersionUID = -234237;
    @Override
    public void doGet(HttpServletRequest request,
            HttpServletResponse response) throws ServletException,
             IOException {
        List<Book> books = new ArrayList<Book>();
        Book book1 = new Book("978-0980839616",
                "Java 7: A Beginner's Tutorial",
                BigDecimal.valueOf(45.00));
        Book book2 = new Book("978-0980331608",
                "Struts 2 Design and Programming: A Tutorial",
```

```
                BigDecimal.valueOf(49.95));
        Book book3 = new Book("978-0975212820",
                "Dimensional Data Warehousing with MySQL: A Tutorial",
                BigDecimal.valueOf(39.95));
        books.add(book1);
        books.add(book2);
        books.add(book3);
        request.setAttribute("books", books);
        RequestDispatcher rd =
                request.getRequestDispatcher("/books.jsp");
        rd.forward(request, response);
    }
}
```

清单 9.6 Book 类

```
package domain;
import java.math.BigDecimal;
public class Book {
    private String isbn;
    private String title;
    private BigDecimal price;

    public Book(String isbn, String title, BigDecimal price) {
        this.isbn = isbn;
        this.title = title;
        this.price = price;
    }

    // getters and setters not shown
}
```

清单 9.7 books.jsp 页面

```
<%@ taglib uri="http://java.sun.com/jsp/jstl/core" prefix="c" %>
<!DOCTYPE html>
<html>
<head>
<title>Book List</title>
<style>table, tr, td {
     border: 1px solid brown;
}
</style>
</head>
<body>
```

```
Books in Simple Table
<table>
    <tr>
        <td>ISBN</td>
        <td>Title</td>
    </tr>
    <c:forEach items="${requestScope.books}" var="book">
    <tr>
        <td>${book.isbn}</td>
        <td>${book.title}</td>
    </tr>
    </c:forEach>
</table>
<br/>
Books in Styled Table
<table>
    <tr style="background:#ababff">
        <td>ISBN</td>
        <td>Title</td>
    </tr>
    <c:forEach items="${requestScope.books}" var="book"
            varStatus="status">
        <c:if test="${status.count%2 == 0}">
            <tr style="background:#eeeeff">
        </c:if>
        <c:if test="${status.count%2 != 0}">
            <tr style="background:#dedeff">
        </c:if>
        <td>${book.isbn}</td>
        <td>${book.title}</td>
    </tr>
    </c:forEach>
</table>

<br/>
ISBNs only:
    <c:forEach items="${requestScope.books}" var="book"
            varStatus="status">
        ${book.isbn}<c:if test="${!status.last}">,</c:if>
    </c:forEach>
</body>
</html>
```

注意，books.jsp 页面显示了图书 3 次，第一次是利用没有 varStatus 属性的 forEach 标签。

```
<table>
    <tr>
```

```
        <th>ISBN</th>
        <th>Title</th>
    </tr>
    <c:forEach items="${requestScope books}" var="book">
    <tr>
        <td>${book.isbn}</td>
        <td>${book.title}</td>
    </tr>
    </c:forEach>
</table>
```

第二次是利用有 varStatus 属性的 forEach 标签来显示，这是为了根据偶数行或奇数行来给表格行设计不同的颜色。

```
<table>
    <tr style="background:#ababff">
        <td>ISBN</td>
        <td>Title</td>
    </tr>
    <c:forEach items="${requestScope.books}" var="book"
            varStatus="status">
        <c:if test="${status.count%2 == 0}">
            <tr style="background:#eeeeff">
        </c:if>
        <c:if test="${status.count%2 != 0}">
            <tr style="background:#dedeff">
        </c:if>
        <td>${book.isbn}</td>
        <td>${book.title}</td>
    </tr>
    </c:forEach>
</table>
```

最后一个 forEach 用来展示以逗号分隔的 ISBN，status.last 用来避免在文末有逗号。

```
<c:forEach items="${requestScope.books}" var="book"
        varStatus="status">
    ${book.isbn}<c:if test="${!status.last}">,</c:if>
</c:forEach>
```

利用以下 URL 可以查看以上范例：

```
http://localhost:8080/iterator/books
```

其输出结果与图 9.3 所示的屏幕截图相似。

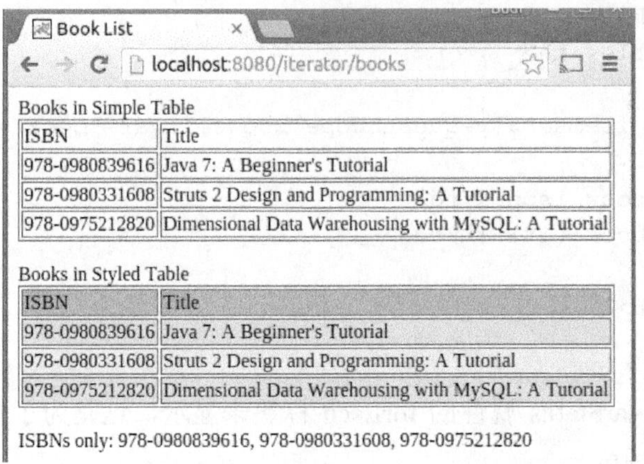

Book List
localhost:8080/iterator/books
Books in Simple Table
ISBN
Title
978-0980839616
Java 7: A Beginner's Tutorial
978-0980331608
Struts 2 Design and Programming: A Tutorial
978-0975212820
Dimensional Data Warehousing with MySQL: A Tutorial
Books in Styled Table
ISBN
Title
978-0980839616
Java 7: A Beginner's Tutorial
978-0980331608
Struts 2 Design and Programming: A Tutorial
978-0975212820
Dimensional Data Warehousing with MySQL: A Tutorial
ISBNs only: 978-0980839616, 978-0980331608, 978-0975212820

图 9.3　对 List 使用 forEach

利用 forEach 还可以遍历 Map。要分别利用 key 和 value 属性引用一个 Map key 和一个 Map 值。遍历 Map 的伪代码如下所示：

```
<c:forEach var="mapItem" items="map">
    ${mapItem.key} : ${mapItem.value}
</c:forEach>
```

清单 9.8 展示了 forEach 与 Map 的结合使用。清单 9.8 中的 BigCitiesServlet 类将两个 Map 实例化，并为它们赋予键/值对。第一个 Map 中的每一个元素都是一个 String/String 对，第二个 Map 中的每一个元素则都是一个 String/String[]对。

清单 9.8　BigCitiesServlet 类

```
package servlet;
import java.io.IOException;
import java.util.HashMap;
import java.util.Map;
import javax.servlet.RequestDispatcher;
import javax.servlet.ServletException;
import javax.servlet.annotation.WebServlet;
import javax.servlet.http.HttpServlet;
import javax.servlet.http.HttpServletRequest;
import javax.servlet.http.HttpServletResponse;

@WebServlet(urlPatterns = {"/bigCities"})
public class BigCitiesServlet extends HttpServlet {
    private static final long serialVersionUID = 1L;

    @Override
```

```
    public void doGet(HttpServletRequest request,
            HttpServletResponse response)
            throws ServletException, IOException {
        Map<String, String> capitals =
                new HashMap<String, String>();
        capitals.put("Indonesia", "Jakarta");
        capitals.put("Malaysia", "Kuala Lumpur");
        capitals.put("Thailand", "Bangkok");
        request.setAttribute("capitals", capitals);
        Map<String, String[]> bigCities =
                new HashMap<String, String[]>();
        bigCities.put("Australia", new String[] {"Sydney",
                "Melbourne", "Perth"});
        bigCities.put("New Zealand", new String[] {"Auckland",
                "Christchurch", "Wellington"});
        bigCities.put("Indonesia", new String[] {"Jakarta",
                "Surabaya", "Medan"});
        request.setAttribute("capitals", capitals);
        request.setAttribute("bigCities", bigCities);
        RequestDispatcher rd =
                request.getRequestDispatcher("/bigCities.jsp");
        rd.forward(request, response);
    }
}
```

在 doGet 方法的结尾处，servlet 跳转到 bigcities.jsp 页面，它利用 forEach 遍历 Map。bigcities.jsp 页面如清单 9.9 所示。

清单 9.9 bigcities.jsp 页面

```
<%@ taglib uri="http://java.sun.com/jsp/jstl/core" prefix="c" %>
<!DOCTYPE html>
<html>
<head>
<title>Big Cities</title>
<style>
table, tr, td {
    border: 1px solid #aaee77;
    padding: 3px;
}
</style>
</head>
<body>
Capitals
<table>
    <tr style="background:#448755;color:white;font-weight:bold">
```

```
            <td>Country</td>
            <td>Capital</td>
        </tr>
        <c:forEach items="${requestScope.capitals}" var="mapItem">
        <tr>
            <td>${mapItem.key}</td>
            <td>${mapItem.value}</td>
        </tr>
        </c:forEach>
    </table>
    <br/>
    Big Cities
    <table>
        <tr style="background:#448755;color:white;font-weight:bold">
            <td>Country</td>
            <td>Cities</td>
        </tr>
        <c:forEach items="${requestScope.bigCities}" var="mapItem">
        <tr>
            <td>${mapItem.key}</td>
            <td>
                <c:forEach items="${mapItem.value}" var="city"
                           varStatus="status">
                    ${city}<c:if test="${!status.last}">,</c:if>
                </c:forEach>
            </td>
        </tr>
        </c:forEach>
    </table>
    </body>
    </html>
```

最重要的是，第二个 forEach 中还嵌套了另一个 forEach：

```
<c:forEach items="${requestScope.bigCities}" var="mapItem">
        <c:forEach items="${mapItem.value}" var="city"
                   varStatus="status">
            ${city}<c:if test="${!status.last}">,</c:if>
        </c:forEach>
</c:forEach>
```

这里的第二个 forEach 是遍历 Map 的元素值，它是一个 String 数组。

登录以下网站可以查看到以上范例：

```
http://localhost:8080/iterator/bigCities
```

打开网页后，浏览器上应该会以 HTML 表格的形式，显示出几个国家的首都和大城市，如图 9.4 所示。

图 9.4　对 Map 使用 forEach

9.5.2　forTokens 标签

forTokens 标签用于遍历以特定分隔符隔开的令牌，其语法如下：

```
<c:forTokens items="stringOfTokens" delims="delimiters"
        [var="varName"] [varStatus="varStatusName"]
        [begin="begin"] [end="end"] [step="step"]
>
    body content
</c:forTokens>
```

body content 是 JSP。forTokens 标签的属性见表 9.7。

表 9.7　forTokens 标签的属性

属性	类型	描述
var	字符串	引用遍历的当前项目的有界变量名称
items+	支持的任意类型	要遍历的 token 字符串
varStatus	字符串	保存遍历状态的有界变量名称。类型值为 javax.servlet.jsp.jstl.core.LoopTagStatus
begin+	整数	遍历的起始索引，此处索引是从 0 开始的。如有指定，begin 的值必须大于或者等于 0
end+	整数	遍历的终止索引，此处索引是从 0 开始的
step+	整数	遍历将以 step 指定的间隔的 token 来处理字符串，从第一个 token 开始。如有指定，step 的值必须大于或者等于 1
delims+	字符串	一组分隔符

下面是一个 forTokens 范例：

```
<c:forTokens var="item" items="Argentina,Brazil,Chile" delims=",">
    <c:out value="${item}"/><br/>
</c:forTokens>
```

当将以上 forTokens 粘贴到 JSP 中时，它将会产生如下结果：

```
Argentina
Brazil
Chile
```

9.6 格式化行为

JSTL 提供了格式化和解析数字与日期的标签，它们是 formatNumber、formatDate、timeZone、setTimeZone、parseNumber 和 parseDate。

9.6.1 formatNumber 标签

formatNumber 用于格式化数字。你可以根据需要，利用这个标签的各种属性来获得自己想要的格式。formatNumber 的语法有两种形式。第一种形式没有 body content：

```
<fmt:formatNumber value="numericValue"
        [type="{number|currency|percent}"]
        [pattern="customPattern"]
        [currencyCode="currencyCode"]
        [currencySymbol="currencySymbol"]
        [groupingUsed="{true|false}"]
        [maxIntegerDigits="maxIntegerDigits"]
        [minIntegerDigits="minIntegerDigits"]
        [maxFractionDigits="maxFractionDigits"]
        [minFractionDigits="minFractionDigits"]
        [var="varName"]
        [scope="{page|request|session|application}"]
/>
```

第二种形式有 body content：

```
<fmt:formatNumber [type="{number|currency|percent}"]
        [pattern="customPattern"]
        [currencyCode="currencyCode"]
        [currencySymbol="currencySymbol"]
        [groupingUsed="{true|false}"]
        [maxIntegerDigits="maxIntegerDigits"]
```

```
        [minIntegerDigits="minIntegerDigits"]
        [maxFractionDigits="maxFractionDigits"]
        [minFractionDigits="minFractionDigits"]
        [var="varName"]
        [scope="{page|request|session|application}"]>
    numeric value to be formatted
</fmt:formatNumber>
```

body content 是 JSP。formatNumber 标签的属性见表 9.8。

表 9.8 formatNumber 标签的属性

属性	类型	描述
value+	字符串或数字	要格式化的数值
type+	字符串	说明该值是要被格式化成数字、货币，还是百分比。这个属性值如下：number、currency 和 percent
pattern+	字符串	定制格式化样式
currencyCode+	字符串	ISO 4217 码，见表 9.9
currencySymbol+	字符串	货币符号
groupingUsed+	布尔	说明输出结果中是否包含组分隔符
maxIntegerDigits+	整数	规定输出结果的整数部分最多几位数字
minIntegerDigits+	整数	规定输出结果的整数部分最少几位数字
maxFractionDigits+	整数	规定输出结果的小数部分最多几位数字
minFractionDigits+	整数	规定输出结果的小数部分最少几位数字
var	字符串	将输出结果存为字符串的有界变量名称
scope	字符串	var 的范围。如果有 scope 属性，则必须指定 var 属性

formatNumber 标签的用途之一就是将数字格式化成货币。为此，可以利用 currencyCode 属性来定义一个 ISO 4217 货币代码。部分 ISO 4217 货币代码见表 9.9。

表 9.9 部分 ISO 4217 货币代码

币别	ISO 4217 码	大单位名称	小单位名称
加拿大元	CAD	加元	分
人民币	CNY	元	角
欧元	EUR	欧元	分
日元	JPY	日元	钱
英镑	GBP	英镑	便士
美元	USD	美元	美分

FormatNumber 标签的用法范例见表 9.10。

表 9.10　formatNumber 标签的用法范例

行为	结果
<fmt:formatNumber value="12" type="number"/>	12
<fmt:formatNumber value="12" type="number" minIntegerDigits="3"/>	012
<fmt:formatNumber value="12" type="number" minFractionDigits="2"/>	12.00
<fmt:formatNumber value="123456.78" pattern=".000"/>	123456.780
<fmt:formatNumber value="123456.78" pattern="#,#00.0#"/>	123,456.78
<fmt:formatNumber value="12" type="currency"/>	$12.00
<fmt:formatNumber value="12" type="currency" currencyCode="GBP"/>	GBP 12.00
<fmt:formatNumber value="0.12" type="percent"/>	12%
<fmt:formatNumber value="0.125" type="percent" minFractionDigits="2"/>	12.50%

注意，在格式化货币时，如果没有定义 currencyCode 属性，就无法使用浏览器的本地化。

9.6.2　formatDate 标签

formatDate 标签用于格式化日期，其语法如下：

```
<fmt:formatDate value="date"
        [type="{time|date|both}"]
        [dateStyle="{default|short|medium|long|full}"]
        [timeStyle="{default|short|medium|long|full}"]
        [pattern="customPattern"]
        [timeZone="timeZone"]
        [var="varName"]
        [scope="{page|request|session|application}"]
/>
```

body content 为 JSP。formatDate 标签的属性见表 9.11。

表 9.11　formatDate 标签的属性

属性	类型	描述
value+	java.util.Date	要格式化的日期和/或时间
type+	字符串	说明要格式化的是时间、日期，还是时间与日期部分都格式化

续表

属性	类型	描述
dataStyle+	字符串	预定义日期的格式化样式，遵循 java.text.DateFormat 中定义的语义
timeStyle+	字符串	预定义时间的格式化样式，遵循 java.text.DateFormat 中定义的语义
pattern+	字符串	定制格式化样式
timezone+	字符串或 java.util.TimeZone	定义用于显示时间的时区
var	字符串	将输出结果存储为字符串的有界变量名称
scope	字符串	var 的范围

timeZone 属性的可能值，请查看 9.6.3 小节。

下列代码利用 formatDate 标签格式化有界变量 now 引用的 java.util.Date 对象。

```
Default: <fmt:formatDate value="${now}"/>
Short: <fmt:formatDate value="${now}" dateStyle="short"/>
Medium: <fmt:formatDate value="${now}" dateStyle="medium"/>
Long: <fmt:formatDate value="${now}" dateStyle="long"/>
Full: <fmt:formatDate value="${now}" dateStyle="full"/>
```

下面的 formatDate 标签用于格式化时间：

```
Default: <fmt:formatDate type="time" value="${now}"/>
Short: <fmt:formatDate type="time" value="${now}"
        timeStyle="short"/>
Medium: <fmt:formatDate type="time" value="${now}"
        timeStyle="medium"/>
Long: <fmt:formatDate type="time" value="${now}" timeStyle="long"/>
Full: <fmt:formatDate type="time" value="${now}" timeStyle="full"/>
```

下面的 formatDate 标签用于格式化日期和时间：

```
Default: <fmt:formatDate type="both" value="${now}"/>
Short date short time: <fmt:formatDate type="both"
  value="${now}" dateStyle="short" timeStyle="short"/>
Long date long time format: <fmt:formatDate type="both"
  value="${now}" dateStyle="long" timeStyle="long"/>
```

下面的 formatDate 标签用于格式化带时区的时间：

```
Time zone CT: <fmt:formatDate type="time" value="${now}"
        timeZone="CT"/><br/>
Time zone HST: <fmt:formatDate type="time" value="${now}"
```

```
        timeZone="HST"/><br/>
```

下面的 formatDate 标签利用定制模式来格式化日期和时间：

```
<fmt:formatDate type="both" value="${now}" pattern="dd.MM.yy"/>
<fmt:formatDate type="both" value="${now}" pattern="dd.MM.yyyy"/>
```

9.6.3　timeZone 标签

timeZone 标签用于定义时区，使其 body content 中的时间信息按指定时区进行格式化或者解析。其语法如下：

```
<fmt:timeZone value="timeZone">
    body content
</fmt:timeZone>
```

body content 是 JSP。属性值可以是类型为 String 或者 java.util.TimeZone 的动态值。美国和加拿大时区的值见表 9.12。

如果 value 属性为 null 或者 empty，则使用 GMT 时区。

下面的范例用 timeZone 标签格式化带时区的日期。

```
<fmt:timeZone value="GMT+1:00">
    <fmt:formatDate value="${now}" type="both"
            dateStyle="full" timeStyle="full"/>
</fmt:timeZone>
<fmt:timeZone value="HST">
    <fmt:formatDate value="${now}" type="both"
            dateStyle="full" timeStyle="full"/>
</fmt:timeZone>
<fmt:timeZone value="CST">
    <fmt:formatDate value="${now}" type="both"
            dateStyle="full" timeStyle="full"/>
</fmt:timeZone>
```

表 9.12　美国和加拿大时区的值

缩写	全名	时区
NST	纽芬兰标准时间	UTC-3:30
NDT	纽芬兰夏时制	UTC-2:30
AST	大西洋标准时间	UTC-4
ADT	大西洋夏时制	UTC-3
EST	东部标准时间	UTC-5

续表

缩写	全名	时区
EDT	东部夏时制	UTC-4
ET	东部时间，如 EST 或 EDT	*
CST	中部标准时间	UTC-6
CDT	中部夏时制	UTC-5
CT	中部时间，如 CST 或 CDT	*
MST	山地标准时间	UTC-7
MDT	山地夏时制	UTC-6
MT	山地时间，如 MST 或 MDT	*
PST	太平洋标准时间	UTC-8
PDT	太平洋夏时制	UTC-7
PT	太平洋时间，如 PST 或 PDT	*
AKST	阿拉斯加标准时间	UTC-9
AKDT	阿拉斯加夏时制	UTC-8
HST	夏威夷标准时间	UTC-10

9.6.4 setTimeZone 标签

setTimeZone 标签用于将指定时区保存在一个有界变量或者时间配置变量中。setTimeZone 的语法如下：

```
<fmt:setTimeZone value="timeZone" [var="varName"]
        [scope="{page|request|session|application}"]
/>
```

表 9.13 展示了 setTimeZone 标签的属性。

表 9.13 setTimeZone 标签的属性

属性	类型	描述
value+	字符串或 java.util.TimeZone	时区
var	字符串	保存类型为 java.util.TimeZone 的时区的有界变量
scope	字符串	var 的范围或者时区配置变量

9.6.5 parseNumber 标签

parseNumber 标签用于将以字符串表示的数字、货币或者百分比解析成数字。其语法有两种形式。第一种形式没有 body content：

```
<fmt:parseNumber value="numericValue"
        [type="{number|currency|percent}"]
        [pattern="customPattern"]
        [parseLocale="parseLocale"]
        [integerOnly="{true|false}"]
        [var="varName"]
        [scope="{page|request|session|application}"]
/>
```

第二种形式有 body content：

```
<fmt:parseNumber [type="{number|currency|percent}"]
        [pattern="customPattern"]
        [parseLocale="parseLocale"]
        [integerOnly="{true|false}"]
        [var="varName"]
        [scope="{page|request|session|application}"]>
    numeric value to be parsed
</fmt:parseNumber>
```

body content 是 JSP。parseNumber 标签的属性见表 9.14。

下面的 parseNumber 标签就是解析有界变量 quantity 引用的值，并将结果保存在有界变量 formattedNumber 中。

```
<fmt:parseNumber var="formattedNumber" type="number"
        value="${quantity}"/>
```

表 9.14　parseNumber 标签的属性

属性	类型	描述
value+	字符串或数字	要解析的字符串
type+	字符串	说明该字符串是要解析成数字、货币，还是百分比
pattern+	字符串	定制格式化样式，决定 value 属性中的字符串要如何解析
parseLocale+	字符串或者 java.util.Locale	定义 locale，在解析操作期间将其默认格式化样式或将 pattern 属性定义的样式应用其中
integerOnly+	布尔	说明是否只解析指定值的整数部分
var	字符串	保存输出结果的有界变量名称
scope	字符串	var 的范围

9.6.6 parseDate 标签

parseDate 标签以区分地域的格式解析以字符串表示的日期和时间。其语法有两种形式。第一种形式没有 body content：

```
<fmt:parseDate value="dateString"
        [type="{time|date|both}"]
        [dateStyle="{default|short|medium|long|full}"]
        [timeStyle="{default|short|medium|long|full}"]
        [pattern="customPattern"]
        [timeZone="timeZone"]
        [parseLocale="parseLocale"]
        [var="varName"]
        [scope="{page|request|session|application}"]
/>
```

第二种形式有 body content：

```
<fmt:parseDate [type="{time|date|both}"]
        [dateStyle="{default|short|medium|long|full}"]
        [timeStyle="{default|short|medium|long|full}"]
        [pattern="customPattern"]
        [timeZone="timeZone"]
        [parseLocale="parseLocale"]
        [var="varName"]
        [scope="{page|request|session|application}"]>
    date value to be parsed
</fmt:parseDate>
```

body content 是 JSP。表 9.15 列出了 parseDate 标签的属性。

表 9.15　parseDate 标签的属性

属性	类型	描述
value+	字符串	要解析的字符串
type+	字符串	说明要解析的字符串中是否包含日期、时间或二者均有
dateStyle+	字符串	日期的格式化样式
timeStyle+	字符串	时间的格式化样式
pattern+	字符串	定制格式化样式，决定要如何解析该字符串
timeZone+	字符串或者 java.util.TimeZone	定义时区，使日期字符串中的时间信息均根据它来解析
parseLocale+	字符串或者 java.util.Locale	定义 locale，在解析操作期间用其默认格式化样式，或将 pattern 属性定义的样式应用其中
var	字符串	保存输出结果的有界变量名称
scope	字符串	var 的范围

下面的 parseDate 标签用于解析有界变量 myDate 引用的日期，并将得到的 java.util.Date 保存在一个页面范围的有界变量 formattedDate 中。

```
<c:set var="myDate" value="12/12/2005"/>
<fmt:parseDate var="formattedDate" type="date"
        dateStyle="short" value="${myDate}"/>
```

9.7 函数

除了定制行为外，JSTL 1.1 和 JSTL 1.2 还定义了一套可以在 EL 表达式中使用的标准函数。这些函数都集中放在 function 标签库中。要使用这些函数，必须在 JSP 的最前面使用以下的 taglib 指令：

```
<%@ taglib uri="http://java.sun.com/jsp/jstl/functions"
        prefix="fn" %>
```

调用函数时，要以下列格式使用一个 EL：

```
${fn:functionName}
```

这里的 functionName 是函数名。

大部分函数都用于字符串操作。例如，length 函数用于字符串和集合，并返回集合或者数组中的项数，或者返回一个字符串的字符数。

9.7.1 contains 函数

contains 函数用于测试一个字符串中是否包含指定的子字符串。如果字符串中包含该子字符串，则返回值为 True，否则，返回 False。其语法如下：

```
contains(string, substring).
```

例如，下面这两个 EL 表达式都将返回 True：

```
<c:set var="myString" value="Hello World"/>
${fn:contains(myString, "Hello")}

${fn:contains("Stella Cadente", "Cadente")}
```

9.7.2 containsIgnoreCase 函数

containsIgnoreCase 函数与 contains 函数相似，但测试是区分大小写的，其语法如下：

```
containsIgnoreCase(string, substring)
```

例如，下列的 EL 表达式将返回 True：

```
${fn:containsIgnoreCase("Stella Cadente", "CADENTE")}
```

9.7.3 endsWith 函数

endsWith 函数用于测试一个字符串是否以指定的后缀结尾。其返回值是一个 Boolean，语法如下：

```
endsWith(string, suffix)
```

例如，下列的 EL 表达式将返回 True：

```
${fn:endsWith("Hello World", "World")}
```

9.7.4 escapeXml 函数

escapeXml 函数用于给 String 编码。这种转换与 out 标签将其 escapeXml 属性设为 True 一样。escapeXml 的语法如下：

```
escapeXml(string)
```

例如，下面的 EL 表达式：

```
${fn:escapeXml("Use <br/> to change lines")}
```

将被渲染成：

```
Use &lt;br/&gt; to change lines
```

9.7.5 indexOf 函数

indexOf 函数返回指定子字符串在某个字符串中第一次出现时的索引。如果没有找到指定的子字符串，则返回-1。其语法如下：

```
indexOf(string, substring)
```

例如，下面的 EL 表达式将返回 7:

```
${fn:indexOf("Stella Cadente", "Cadente")}
```

9.7.6 join 函数

join 函数将一个 String 数组中的所有元素都合并成一个字符串，并用指定的分隔符分开，其语法如下：

```
join(array, separator)
```

如果这个数组为 null，就会返回一个空字符串。

例如，如果 myArray 是一个 String 数组，它带有两个元素“my”和“world”，那么，下面的 EL 表达式：

```
${fn:join(myArray,",")}
```

将返回“my, world”。

9.7.7 length 函数

length 函数用于返回集合中的项数，或者字符串中的字符数，其语法如下：

```
length{input}
```

下面的 EL 表达式将返回 14。

```
${fn:length("Stella Cadente", "Cadente")}
```

9.7.8 replace 函数

replace 函数将字符串中出现的所有 beforeString 都用 afterString 替换，并返回结果，其语法如下：

```
replace(string, beforeSubstring, afterSubstring)
```

例如，下面的 EL 表达式：

```
${fn:replace("Stella Cadente", "e", "E")}
```

将返回“StElla CadEntE”。

9.7.9 split 函数

split 函数用于将一个字符串分割成一个子字符串数组。它的作用与 join 相反。例如，下列代码分割字符串“my, world”，并将结果保存在有界变量 split 中。随后，利用 forEach 标签将 split 格式化成一个 HTML 表。

```
<c:set var="split" value='${fn:split("my,world",",")}'/>
<table>
<c:forEach var="substring" items="${split}">
    <tr><td>${substring}</td></tr>
</c:forEach>
</table>
```

结果为：

```
<table>
    <tr><td>my</td></tr>
    <tr><td>world</td></tr>
</table>
```

9.7.10 startsWith 函数

startsWith 函数用于测试一个字符串是否以指定的前缀开头，其语法如下：

```
startsWith(string, prefix)
```

例如，下面的 EL 表达式将返回 True：

```
${fn:startsWith("Stella Cadente", "St")}
```

9.7.11 substring 函数

substring 函数用于返回一个从指定基于 0 的起始索引（含）到指定基于 0 的终止索引的子字符串，其语法如下：

```
substring(string, beginIndex, endIndex)
```

下面的 EL 表达式将返回“Stel”。

```
${fn:substring("Stella Cadente", 0, 4)}
```

9.7.12 substringAfter 函数

substringAfter 函数用于返回指定子字符串第一次出现后的字符串部分，其语法如下：

```
substringAfter(string, substring)
```

例如，下面的 EL 表达式：

```
${fn:substringAfter("Stella Cadente", "e")}
```

将返回“lla Cadente”。

9.7.13 substringBefore 函数

substringBefore 函数用于返回指定子字符串第一次出现前的字符串部分，其语法如下：

```
substringBefore(string, substring)
```

例如，下面的 EL 表达式将返回“St”。

```
${fn:substringBefore("Stella Cadente", "e")}
```

9.7.14　toLowerCase 函数

toLowerCase 函数将一个字符串转换成它的小写版本，其语法如下：

```
toLowerCase(string)
```

例如，下面的 EL 表达式将返回“stella cadente”：

```
${fn:toLowerCase("Stella Cadente")}
```

9.7.15　toUpperCase 函数

toUpperCase 函数将一个字符串转换成它的大写版本，其语法如下：

```
toUpperCase(string)
```

例如，下面的 EL 表达式将返回“STELLA CADENTE”。

```
${fn:toUpperCase("Stella Cadente")}
```

9.7.16　trim 函数

trim 函数用于删除一个字符串开头和结尾的空白，其语法如下：

```
trim(string)
```

例如，下面的 EL 表达式将返回“Stella Cadente”。

```
${fn:trim("           Stella Cadente   ")}
```

9.8　小结

JSTL 可以完成一般的任务（如遍历、集合和条件）、处理 XML 文档、格式化文本、访问数据库以及操作数据，等等。本章介绍了比较重要的一些标签，如操作有界对象的标签（out、set、remove），执行条件测试的标签（if、choose、when、otherwise），遍历集合或 token 的标签（forEach、forTokens），解析和格式化日期与数字的标签（parseNumber、formatNumber、parseDate、formatDate 等），以及可以在 EL 表达式中使用的 JSTL 1.2 函数。

第 10 章 国际化

在这个全球化的时代，现在比过去更需要能够编写可以在讲不同语言的国家和地区部署的应用程序。在这方面，需要了解两个术语。第一个术语是国际化，常常缩写为 i18n，因为其单词 internationalization 以 i 开头，以 n 结尾，在它们之间有 18 个字母。国际化是开发支持多语言和数据格式的应用程序的技术，无需重写编程逻辑。

第二个术语是本地化，这是将国际化应用程序改成支持特定语言区域（locale）的技术。语言区域是指一个特定的地理、政治或者文化区域。要考虑到语言区域的一个操作，就称作区分语言区域的操作。例如，显示日期就是一个区分语言区域的操作，因为日期必须以用户所在的国家或者地区使用的格式显示。2016 年 11 月 15 日，在美国显示为 11/15/2016，但在澳大利亚则显示为 15/11/2016。与国际化缩写为 i18n 一样，本地化缩写为 l10n。

Java 谨记国际化的需求，为字符和字符串提供了 Unicode 支持。因此，用 Java 编写国际化的应用程序是一件很容易的事情。国际化应用程序的具体方式取决于有多少静态数据需要以不同的语言显示出来。这里有两种方法：

（1）如果大量数据是静态的，就要针对每一个语言区域单独创建一个资源版本。这种方法一般适用于带有大量静态 HTML 页面的 Web 应用程序。这个很简单，不在本章讨论范围。

（2）如果需要国际化的静态数据量有限，就可以将文本元素，如元件标签和错误消息隔离为文本文件。每个文本文件中都保存着一个语言区域的所有文本元素译文。随后，应用程序会自动获取每一个元素。这样做的优势是显而易见的。每个文本元素无需重新编译应用程序，便可轻松地进行编辑。这正是本章要讨论的技术。

本章将首先解释什么是语言区域，接着讲解国际化应用程序技术，最后介绍一个 Spring MVC 范例。

10.1　语言区域

Java.util.Locale 类表示一个语言区域。一个 Locale 对象包含 3 个主要元件：language、country 和 variant。Language 无疑是最重要的部分；但是，语言本身有时并不足以区分一个语言区域。例如，讲英语的国家有很多，如美国和英国。但是，在美国讲的英语，与在英国用的英语并非一模一样。因此，必须指定语言国家。再举一个例子，在中国大陆使用的汉语，与在台湾地区用的汉语也是不完全一样的。

参数 variant 是一个特定于供应商或者特定于浏览器的代号。例如，用 WIN 表示 Windows，用 MAC 表示 Macintosh，用 POSIX 表示 POSIX。两个 variant 之间用一个下划线隔开，并将最重要的部分放在最前面。例如，传统西班牙语，用 language、country 和 variant 参数构造一个 locale 分别是 es，ES，Traditional_WIN。

构造 Locale 对象时，要使用 Locale 类的其中一个构造器：

```
public Locale(java.lang.String language)
public Locale(java.lang.String language, java.lang.String country)
public Locale(java.lang.String language, java.lang.String country,
        java.lang.String variant)
```

语言代号是一个有效的 ISO 语言码。表 10.1 显示了 ISO 639 语言码范例。

表 10.1　ISO 639 语言码范例

代码	语言
de	德语
el	希腊语
en	英语
es	西班牙语
fr	法语
hi	印地语
it	意大利语
ja	日语
nl	荷兰语
pt	葡萄牙语
ru	俄语
zh	汉语

参数 country 是一个有效的 ISO 国家码，由两个字母组成，ISO 3166（http://userpage.chemie.fuberlin.de/diverse/doc/ISO_3166.html）中指定为大写字母。表 10.2 展示了 ISO 3166 国家码范例。

表 10.2　ISO 3166 国家码范例

国家	代码
澳大利亚	AU
巴西	BR
加拿大	CA
中国	CN
埃及	EG
法国	FR
德国	DE
印度	IN
墨西哥	MX
瑞士	CH
英国	GB
美国	US

例如，要构造一个表示加拿大所用英语的 Locale 对象，可以像下面这样编写：

```
Locale locale = new Locale("en", "CA");
```

此外，Locale 类提供了 static final 域，用来返回特定国家或语言的语言区域，如 CANADA_FRENCH、CHINA、CHINESE、ENGLISH、FRANCE、FRENCH、UK、US 等。因此，也可以通过调用其 static 域来构造 Locale 对象。

```
Locale locale = Locale.CANADA_FRENCH;
```

此外，静态的 getDefault 方法会返回用户计算机的语言区域。

```
Locale locale = Locale.getDefault();
```

10.2　国际化 Spring MVC 应用程序

国际化和本地化应用程序时，需要具备以下条件：

（1）将文本组件隔离成属性文件。

（2）选择和读取正确的属性文件。

下面详细介绍这两个步骤，并进行简单的示范。

10.2.1　将文本组件隔离成属性文件

国际化的应用程序是将每一个语言区域的文本元素都单独保存在一个独立的属性文件中。每个文件中都包含 key/value 对，并且每个 key 都唯一表示一个特定语言区域的对象。key 始终是字符串，value 则可以是字符串，也可以是其他任意类型的对象。例如，为了支持美国英语、德语以及汉语，就要有 3 个属性文件，它们都有着相同的 key。

以下是英语版的属性文件。注意，它有 greetings 和 farewell 两个 key。

```
greetings = Hello
farewell = Goodbye
```

德国版的属性文件如下：

```
greetings = Hallo
farewell = Tschüß
```

汉语版的属性文件如下：

```
greetings=\u4f60\u597d
farewell=\u518d\u89c1
```

如果你是中文用户，你可以使用任何中文文本编辑器并写入汉字字符。 完成后，将文件转换为 Unicode。

现在，要学习 java.util.ResourceBundle 类。它能使你轻松地选择和读取特定于用户语言区域的属性，以及查找值。ResourceBundle 是一个抽象类，但它提供了静态的 getBundle 方法，返回一个具体子类的实例。

ResourceBundle 有一个基准名，它可以是任意名称。但是，为了让 ResourceBundle 正确地选择属性文件，这个文件名中最好必须包含基准名 ResourceBundle，后面再接下划线、语言码，还可以选择再加一条下划线和国家码。属性文件名的格式如下所示：

```
basename_languageCode_countryCode
```

例如，假设基准名为 MyResources，并且定义了以下 3 个语言区域：

- US-en;

- DE-de;
- CN-zh。

那么，就会得到下面这 3 个属性文件：

- MyResources_en_US.properties;
- MyResources_de_DE.properties;
- MyResources_zh_CN.properties。

10.2.2 选择和读取正确的属性文件

如前所述，ResourceBundle 是一个抽象类。尽管如此，还是可以通过调用它的静态 getBundle 方法来获得一个 ResourceBundle 实例。它的重载签名如下：

```
public static ResourceBundle getBundle(java.lang.String baseName)
public static ResourceBundle getBundle(java.lang.String baseName,
        Locale locale)
```

例如：

```
ResourceBundle rb = ResourceBundle.getBundle("MyResources", Locale.US);
```

这样将会加载 ResourceBundle 在相应属性文件中的值。

如果没有找到合适的属性文件，ResourceBundle 对象就会返回到默认的属性文件。默认属性文件的名称为基准名加一个扩展名 properties。在这个例子中，默认文件就是 MyResources.properties。如果没有找到默认文件，将抛出 java.util.MissingResourceException。

随后，读取值，利用 ResourceBundle 类的 getString 方法传入一个 key。

```
public java.lang.String getString(java.lang.String key)
```

如果没有找到指定 key 的入口，将会抛出 java.util.MissingResourceException。

在 Spring MVC 中，不直接使用 ResourceBundle，而是利用 messageSource bean 告诉 Spring MVC 要将属性文件保存在哪里。例如，下面的 messageSource bean 读取了两个属性文件。

```
<bean id="messageSource" class="org.springframework.context.support.
ReloadableResourceBundleMessageSource">
    <property name="basenames" >
        <list>
            <value>resource/messages</value>
```

```
            <value>resource/labels</value>
        </list>
    </property>
</bean>
```

上面的 bean 定义中用 ReloadableResourceBundleMessageSource 类作为实现。另一个实现中包含了 ResourceBundleMessageSource，它是不能重新加载的。这意味着，如果在任意属性文件中修改了某一个属性 key 或者 value，并且正在使用 ResourceBundleMessageSource，那么要使修改生效，就必须先重启 JVM。另一方面，也可以将 ReloadableResourceBundleMessageSource 设为可重新加载。

这两个实现之间的另一个区别是：使用 ReloadableResourceBundleMessageSource 时，是在应用程序目录下搜索这些属性文件。而使用 ResourceBundleMessageSource 时，属性文件则必须放在类路径下，即 WEB-INF/class 目录下。

还要注意，如果只有一组属性文件，则可以用 basename 属性代替 basenames，像下面这样：

```
<bean id="messageSource" class="org.springframework.context.support.
➥ ResourceBundleMessageSource">
    <property name="basename" value="resource/messages"/>
</bean>
```

10.3　告诉 Spring MVC 使用哪个语言区域

为用户选择语言区域时，最常用的方法或许是读取用户浏览器的 accept-language 标题值。accept-language 标题提供了关于用户偏好哪种语言的信息。

选择语言区域的其他方法还包括读取某个 session 属性或者 cookie。

在 Spring MVC 中选择语言区域，可以使用语言区域解析器 bean。它有几个实现，包括：

- AcceptHeaderLocaleResolver；
- SessionLocaleResolver；
- CookieLocaleResolver。

所有这些实现都是 org.springframework.web.servlet.i18n 包的组成部分。AcceptHeaderLocaleResolver 或许是其中最容易使用的一个。如果选择使用这个语言区域解析器，Spring MVC 将会读取浏览器的 accept-language 标题，来确定浏览器要接受哪个（些）语言区域。如果浏览器的某个语言区域与 Spring MVC 应用程序支持的某个语言区域匹配，就会使用这个语言区域。如果没有找到匹配的语言区域，则使用默认的语言区域。

下面是使用 AcceptHeaderLocaleResolver 的 localeResolver bean 定义：

```
<bean id="localeResolver" class="org.springframework.web.servlet.i18n.
➥ AcceptHeaderLocaleResolver">
</bean>
```

10.4 使用 message 标签

在 Spring MVC 中显示本地化消息的最容易方法是使用 Spring 的 message 标签。为了使用这个标签，要在使用该标签的所有 JSP 页面最前面声明如下的 taglib 指令。

```
<%@taglib prefix="spring"
    uri="http://www.springframework.org/tags"%>
```

message 标签的属性见表 10.3。所有这些属性都是可选的。

表 10.3 message 标签的属性

属性	描述
arguments	该标签的参数写成一个有界的字符串、一个对象数组或者单个对象
argumentSeparator	用来分隔该标签参数的字符
code	获取消息的 key
htmlEscape	接受 True 或者 False，表示被渲染文本是否应该进行 HTML 转义
javaScriptEscape	接受 True 或者 False，表示被渲染文本是否应该进行 JavaScript 转义
message	MessageSourceResolvable 参数
scope	保存 var 属性中定义的变量的范围
text	如果 code 属性不存在，或者指定码无法获取消息，所显示的默认文本
var	用于保存消息的有界变量

10.5 范例

举例来说，i18n 应用程序展示了用 localeResolver bean 将 JSP 页面中的消息本地化的方法。其目录结构如图 10.1 所示，i18n 的 Spring MVC 配置文件如清单 10.1 所示。

```
webapp
  css
    main.css
  WEB-INF
    classes
    config
      springmvc-config.xml
    jsp
      ProductDetails.jsp
      ProductForm.jsp
    lib
    resource
      labels_fr.properties
      labels.properties
      messages_en.properties
      messages_fr.properties
      messages.properties
    web.xml
```

图 10.1　i18n 的目录结构

清单 10.1　i18n 的 Spring MVC 配置文件

```
<?xml version="1.0" encoding="UTF-8"?>
<beans xmlns="http://www.springframework.org/schema/beans"
    xmlns:xsi="http://www.w3.org/2001/XMLSchema-instance"
    xmlns:p="http://www.springframework.org/schema/p"
    xmlns:mvc="http://www.springframework.org/schema/mvc"
    xmlns:context="http://www.springframework.org/schema/context"
    xsi:schemaLocation="
        http://www.springframework.org/schema/beans
        http://www.springframework.org/schema/beans/spring-beans.xsd
        http://www.springframework.org/schema/mvc
        http://www.springframework.org/schema/mvc/spring-mvc.xsd
        http://www.springframework.org/schema/context
        http://www.springframework.org/schema/context/springcontext.xsd">

    <context:component-scan base-package="controller" />
    <mvc:annotation-driven/>

    <mvc:resources mapping="/css/**" location="/css/" />
    <mvc:resources mapping="/*.html" location="/" />

    <bean id="viewResolver" class="org.springframework.web.servlet.view.
➥ InternalResourceViewResolver">
        <property name="prefix" value="/WEB-INF/jsp/" />
        <property name="suffix" value=".jsp" />
    </bean>
```

```
    <bean id="messageSource" class="org.springframework.context.support.
➥ ReloadableResourceBundleMessageSource">
        <property name="basenames" >
            <list>
                <value>/WEB-INF/resource/messages</value>
                <value>/WEB-INF/resource/labels</value>
            </list>
        </property>
    </bean>

    <bean id="localeResolver" class="org.springframework.web.servlet.i18n.
➥ AcceptHeaderLocaleResolver">
    </bean>
</beans>
```

这里用到了 messageSource bean 和 localeResolver bean 这两个 bean。messageSource bean 声明用两个基准名设置了 basenames 属性：/WEB-INF/resource/messages 和/WEB-INF/resource/labels。localeResolver bean 利用 AcceptHeaderLocaleResolver 类实现消息的本地化。

它支持 en 和 fr 两个语言区域，因此每个属性文件都有两种版本。为了实现本地化，JSP 页面中的每一段文本都要用 message 标签代替。清单 10.2 展示了 ProductForm.jsp 页面。注意，为了达到调试的目的，当前的语言区域和 accept-language 标题显示在页面的最前面。

清单 10.2　ProductForm.jsp 页面

```
<%@ taglib prefix="form"
    uri="http://www.springframework.org/tags/form"%>
<%@ taglib
    prefix="spring" uri="http://www.springframework.org/tags"%>
<%@ taglib uri="http://java.sun.com/jsp/jstl/core" prefix="c"%>
<!DOCTYPE HTML>
<html>
<head>
<title><spring:message code="page.productform.title"/></title>
<style type="text/css">@import url("<c:url
    value="/css/main.css"/>");</style>
</head>
<body>
<div id="global">
Current Locale : ${pageContext.response.locale}
<br/>
accept-language header: ${header["accept-language"]}
```

```
<form:form commandName="product" action="product_save"
    method="post">
    <fieldset>
        <legend><spring:message code="form.name"/></legend>
        <p>
            <label for="name"><spring:message
                code="label.productName" text="default text" />:
            </label>
            <form:input id="name" path="name"
                cssErrorClass="error"/>
            <form:errors path="name" cssClass="error"/>
        </p>
        <p>
            <label for="description"><spring:message
                code="label.description "/>
            </label>
            <form:input id="description" path="description"/>
        </p>
        <p>
            <label for="price"><spring:message code="label.price"
               text="default text" />: </label>
            <form:input id="price" path="price"
                cssErrorClass="error"/>
            <form:errors path="price" cssClass="error"/>
        </p>
        <p id="buttons">
            <input id="reset" type="reset" tabindex="4"
                value="<spring:message code="button.reset"/>">
            <input id="submit" type="submit" tabindex="5"
                value="<spring:message code="button.submit"/>">
        </p>
    </fieldset>
</form:form>
</div>
</body>
</html>
```

为了测试 i18n 的国际化特性，要修改浏览器的 accept-language 标签。

对于 Chrome 浏览器，打开“设置”页面，点击“显示高级设置”，点击“语言和输入设置”，添加并移动语言到列表的顶部。

对于 IE 浏览器，到 Tools >Internet Options > General (tab) > Languages > Language Preference 中修改。在 Language Preference 窗口中，单击 Add 按钮添加一种语言。当选择了多种语言时，为了修改某一种语言的优先值，要使用 Move up 和 Move down 按钮。

在其他浏览器中修改 accept-language 标题的说明，请访问以下网址查阅：

```
http://www.w3.org/International/questions/qa-lang-priorities.en.php
```

如果要对这个应用程序进行测试，访问以下 URL：

```
http://localhost:8080/i18n/add-product
```

你将会看到 Product 表的英语版和法语版，分别如图 10.2 和图 10.3 所示。

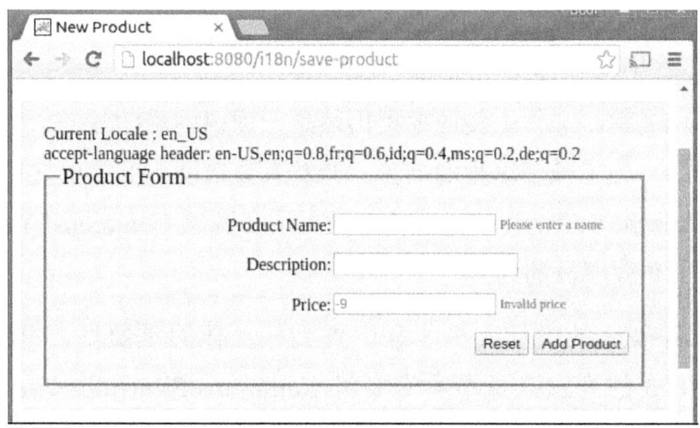

图 10.2　语言区域为 en_US 的 Product 表

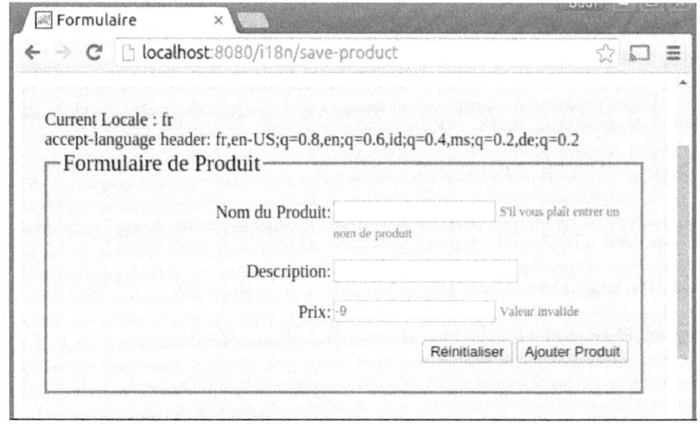

图 10.3　语言区域为 fr_CA 的 Product 表

10.6　小结

本章讲解了如何开发国际化的应用程序，首先介绍了 java.util.Locale 类和 java.util.ResourceBundle 类，然后展示了一个国际化应用程序的例子。

第 11 章
上传文件

Servlet 技术出现之前，文件上传的编程仍然是一项很困难的任务，它涉及在服务器端解析原始的 HTTP 响应。为了减轻编程的痛苦，开发人员借助于商业的文件上传组件。值得庆幸的是，2003 年，Apache Software Foundation 发布了开源的 Commons FileUpload 元件，它很快成为全球 Servlet/JSP 程序员的利器。

经过很多年，Servlet 的设计人员才意识到文件上传的重要性，并终于成为 Servlet 3 的内置特性。Servlet 3 的开发人员不再需要将 Commons FileUpload 元件导入到他们的项目中去。

为此，在 Spring MVC 中处理文件上传有两种方法：

（1）使用 Apache Commons FileUpload 组件。

（2）利用 Servlet 3.0 及其更高版本的内置支持。如果要将应用程序部署到支持 Servlet 3.0 及其更高版本的容器中，则只能使用这种方法。

无论选择哪一种方法，都要利用相同的 API 来处理已经上传的文件。本章将介绍如何在需要支持文件上传的 Spring MVC 应用程序中使用 Commons FileUpload 和 Servlet 3 文件上传特性。此外，本章还将展示如何通过 HTML 5 增强用户体验。

11.1 客户端编程

为了上传文件，必须将 HTML 表格的 enctype 属性值设为 multipart/form-data，像下面这样：

```
<form action="action" enctype="multipart/form-data" method="post">
    Select a file <input type="file" name="fieldName"/>
    <input type="submit" value="Upload"/>
</form>
```

表格中必须包含类型为 file 的一个 input 元素，它会显示成一个按钮，点击时，它会打开一个对话框，用来选择文件。这个表格中也包含了其他的字段类型，如文本区域，或者隐藏字段。

在 HTML 5 之前，如果想要上传多个文件，必须使用多个文件 input 元素。但是，在 HTML 5 中，通过在 input 元素中引入多个 multiple 属性，使得多个文件的上传变得更加简单。在 HTML 5 中编写以下任意一行代码，便可生成一个按钮来选择多个文件：

```
<input type="file" name="fieldName" multiple/>
<input type="file" name="fieldName" multiple="multiple"/>
<input type="file" name="fieldName" multiple=""/>
```

11.2　MultipartFile 接口

在 Spring MVC 中处理已经上传的文件十分容易。上传到 Spring MVC 应用程序中的文件会被包在一个 MultipartFile 对象中。你唯一的任务就是，用类型为 MultipartFile 的属性编写一个 domain 类。

org.springframework.web.multipart.MultipartFile 接口具有以下方法：

```
byte[] getBytes()
```

它以字节数组的形式返回文件的内容。

```
String getContentType()
```

它返回文件的内容类型。

```
InputStream getInputStream()
```

它返回一个 InputStream，从中读取文件的内容。

```
String getName()
```

它以多部分的形式返回参数的名称。

```
String getOriginalFilename()
```

它返回客户端本地驱动器中的初始文件名。

```
long getSize()
```

它以字节为单位，返回文件的大小。

```
boolean isEmpty()
```

它表示被上传的文件是否为空。

```
void transferTo(File destination)
```

它将上传的文件保存到目标目录下。

接下来的范例将讲解如何获取控制器中的已上传文件。

11.3　用 Commons FileUpload 上传文件

只有实现了 Servlet 3.0 及其更高版本规范的 Servlet 容器，才支持文件上传。对版本低于 Servlet 3.0 的容器，则需要 Apache Commons FileUpload 组件，可以从以下网页下载它：

```
http://commons.apache.org/proper/commons-fileupload/
```

这是一个开源项目，因此是免费的，它还提供了源代码。为了让 Commons FileUpload 成功地工作，还需要另一个 Apache Commons 组件：Apache Commons IO。从以下网页可以下载到 Apache Commons IO：

```
http://commons.apache.org/proper/commons-io/
```

因此，需要将两个 JAR 文件复制到应用程序的 WEB-INF/lib 目录下。Commons FileUpload JAR 的名称遵循以下模式：

```
commons-fileupload-x.y.jar
```

这里的 x 是指该软件的大版本，y 是指小版本。例如，本章使用的名称是 commons-fileupload-1.3.jar。

Commons IO JAR 的名称遵循以下模式：

```
commons-io-x.y.jar
```

这里的 x 是指该软件的大版本，y 是指小版本。例如，本章使用的名称是 commons-io-2.4.jar。

此外，还需要在 Spring MVC 配置文件中定义 multipartResolver bean。

```
<bean id="multipartResolver"
        class="org.springframework.web.multipart.commons.
➥ CommonsMultipartResolver">
    <property name="maxUploadSize" value="2000000"/>
</bean>
```

范例 upload1 展示了如何利用 Apache Commons FileUpload 处理已经上传的文件。这个范例在 Servlet 3.0 容器中也是有效的。upload1 有一个 domain 类，即 Product 类，它包含了一个 MultipartFile 对象列表。该例介绍了如何编写一个处理已上传产品图片的控制器。

11.4　Domain 类

清单 11.1 展示了 domain 类 Product。它与前一个例子中的 Product 类相似，只是清单 11.1 中的这个类还具有类型为 List<MultipartFile>的 images 属性。

清单 11.1　经过修改的 Product domain 类

```
package domain;
import java.io.Serializable;
import java.math.BigDecimal;
import java.util.List;
import javax.validation.constraints.NotNull;
import javax.validation.constraints.Size;
import org.springframework.web.multipart.MultipartFile;

public class Product implements Serializable {
    private static final long serialVersionUID = 74458L;

    @NotNull
    @Size(min=1, max=10)
    private String name;

    private String description;
    private BigDeciimal price;
    private List<MultipartFile> images;

    // getters and setters not shown
}
```

11.5　控制器

upload1 中的控制器如清单 11.2 所示。这个类中有 inputProduct 和 saveProduct 两个处理请求的方法。inputProduct 方法向浏览器发出一个产品表单。saveProduct 方法将已上传的图片文件保存到应用程序目录的 image 目录下。

清单 11.2　ProductController 类

```
package controller;
import java.io.File;
import java.io.IOException;
import java.util.ArrayList;
import java.util.List;
import javax.servlet.http.HttpServletRequest;
import org.apache.commons.logging.Log;
import org.apache.commons.logging.LogFactory;
import org.springframework.stereotype.Controller;
import org.springframework.ui.Model;
import org.springframework.validation.BindingResult;
import org.springframework.web.bind.annotation.ModelAttribute;
import org.springframework.web.bind.annotation.RequestMapping;
import org.springframework.web.multipart.MultipartFile;
import domain.Product;

@Controller
public class ProductController {

    private static final Log logger =
        LogFactory.getLog(ProductController.class);

    @RequestMapping(value = "/input-product")
    public String inputProduct(Model model) {
        model.addAttribute("product", new Product());
        return "ProductForm";
    }

    @RequestMapping(value = "/save-product")
    public String saveProduct(HttpServletRequest servletRequest,
            @ModelAttribute Product product,
            BindingResult bindingResult, Model model) {

        List<MultipartFile> files = product.getImages();
        List<String> fileNames = new ArrayList<String>();
        if (null != files && files.size() > 0) {
            for (MultipartFile multipartFile : files) {

                String fileName =
                        multipartFile.getOriginalFilename();
                fileNames.add(fileName);
                File imageFile = new
                       File(servletRequest.getServletContext()
                       .getRealPath("/image"), fileName);
```

```
                try {
                    multipartFile.transferTo(imageFile);
                } catch (IOException e) {
                    e.printStackTrace();
                }
            }
        }
        // save product here
        model.addAttribute("product", product);
        return "ProductDetails";
    }
}
```

如清单 11.2 中的 saveProduct 方法所示，保存已上传文件是一件很轻松的事情，只需要在 MultipartFile 中调用 transferTo 方法。

11.6 配置文件

清单 11.3 展示了 upload1 的 Spring MVC 配置文件。

清单 11.3 upload1 的 Spring MVC 配置文件

```
<?xml version="1.0" encoding="UTF-8"?>
<beans xmlns="http://www.springframework.org/schema/beans"
    xmlns:xsi="http://www.w3.org/2001/XMLSchema-instance"
    xmlns:p="http://www.springframework.org/schema/p"
    xmlns:mvc="http://www.springframework.org/schema/mvc"
    xmlns:context="http://www.springframework.org/schema/context"
    xsi:schemaLocation="
        http://www.springframework.org/schema/beans
        http://www.springframework.org/schema/beans/spring-beans.xsd
        http://www.springframework.org/schema/mvc
        http://www.springframework.org/schema/mvc/spring-mvc.xsd
        http://www.springframework.org/schema/context
        http://www.springframework.org/schema/context/spring-
➥ context.xsd">

    <context:component-scan base-package="controller" />
    <mvc:annotation-driven/>
    <mvc:resources mapping="/css/**" location="/css/" />
    <mvc:resources mapping="/*.html" location="/" />
    <mvc:resources mapping="/image/**" location="/image/" />
```

```
    <bean id="viewResolver"
            class="org.springframework.web.servlet.view.
➥ InternalResourceViewResolver">
        <property name="prefix" value="/WEB-INF/jsp/" />
        <property name="suffix" value=".jsp" />
    </bean>


    <bean id="multipartResolver"
            class="org.springframework.web.multipart.commons.
➥ CommonsMultipartResolver">
    </bean>
</beans>
```

利用 multipartResolver bean 的 maxUploadSize 属性，可以设置能够接受的最大文件容量。如果没有设置这个属性，则没有最大文件容量限制。没有设置文件容量限制，并不意味着可以上传任意大小的文件。上传过大的文件时要花很长的时间，这样会导致服务器超时。为了处理超大文件的问题，可以利用 HTML 5 File API 将文件切片，然后再分别上传这些文件。

11.7 JSP 页面

用于上传图片文件的 ProductForm.jsp 页面如清单 11.4 所示。

清单 11.4　ProductForm.jsp 页面

```
<%@ taglib prefix="form" uri="http://www.springframework.org/tags/form" %>
<%@ taglib uri="http://java.sun.com/jsp/jstl/core" prefix="c" %>
<!DOCTYPE HTML>
<html>
<head>
<title>Add Product Form</title>
<style type="text/css">@import url("<c:url
       value="/css/main.css"/>");</style>
</head>
<body>

<div id="global">
<form:form commandName="product" action="save-product" method="post"
        enctype="multipart/form-data">
    <fieldset>
        <legend>Add a product</legend>
```

```
        <p>
            <label for="name">Product Name: </label>
            <form:input id="name" path="name"
                cssErrorClass="error"/>
            <form:errors path="name" cssClass="error"/>
        </p>
        <p>
            <label for="description">Description: </label>
            <form:input id="description" path="description"/>
        </p>
        <p>
            <label for="price">Price: </label>
            <form:input id="price" path="price"
                cssErrorClass="error"/>
        </p>
        <p>
            <label for="image">Product Image: </label>
            <input type="file" name="images[0]"/>
        </p>
        <p id="buttons">
            <input id="reset" type="reset" tabindex="4">
            <input id="submit" type="submit" tabindex="5"
                value="Add Product">
        </p>
    </fieldset>
</form:form>
</div>
</body>
</html>
```

注意表单中类型为 file 的 input 元素，它将显示为一个按钮，用于选择要上传的文件。

提交 Product 表单，将会调用 save-product 方法。如果这个方法成功地完成，用户将会跳转到清单 11.5 所示的 ProductDetails.jsp 页面。

清单 11.5　ProductDetails.jsp 页面

```
<%@ taglib uri="http://java.sun.com/jsp/jstl/core" prefix="c" %>
<!DOCTYPE HTML>
<html>
<head>
<title>Save Product</title>
<style type="text/css">@import url("<c:url
       value="/css/main.css"/>");</style>
</head>
<body>
<div id="global">
```

```
    <h4>The product has been saved.</h4>
    <p>
        <h5>Details:</h5>
        Product Name: ${product.name}<br/>
        Description: ${product.description}<br/>
        Price: $${product.price}
        <p>Following files are uploaded successfully.</p>
        <ol>
        <c:forEach items="${product.images}" var="image">
            <li>${image.originalFilename}
            <img width="100" src="<c:url value="/image/"/>
            ${image.originalFilename}"/>
            </li>
        </c:forEach>
        </ol>
    </p>
</div>
</body>
</html>
```

ProductDetails.jsp 页面显示出已保存的 Product 的详细信息及其图片。

11.8　应用程序的测试

要测试这个应用程序，在浏览器中打开以下网址：

```
http://localhost:8080/upload1/input-product
```

你将会看到一个如图 11.1 所示的“Add aProduct”表单。试着在其中输入一些产品信息，并选择一个要上传的文件。

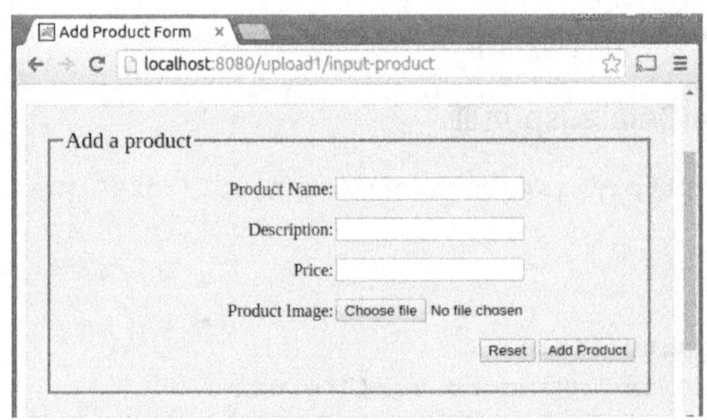

图 11.1　包含一个文件字段的产品表单

单击“Add Product”按钮，就会看到如图 11.2 所示的网页。

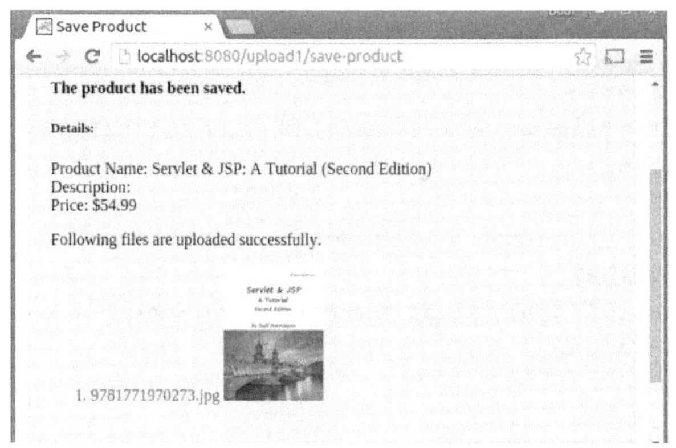

图 11.2　显示已经上传的图片

如果到应用程序目录的 image 目录下查看，就会看到已经上传的图片。

11.9　用 Servlet 3 及其更高版本上传文件

有了 Servlet 3，就不需要 Commons FileUpload 和 Commons IO 元件了。在 Servlet 3 及其以上版本的容器中进行服务器端文件上传的编程，是围绕着注解类型 MultipartConfig 和 javax.servlet.http.Part 接口进行的。处理已上传文件的 Servlets 必须以@MultipartConfig 进行注解。

下列是可能在 MultipartConfig 注解类型中出现的属性，它们都是可选的。

- maxFileSize：上传文件的最大容量，默认值为-1，表示没有限制。大于指定值的文件将会遭到拒绝。
- maxRequestSize：表示多部分 HTTP 请求允许的最大容量，默认值为-1，表示没有限制。
- location：表示在 Part 调用 write 方法时，要将已上传的文件保存到磁盘中的位置。
- fileSizeThreshold：上传文件超出这个容量界限时，会被写入磁盘。

Spring MVC 的 DispatcherServlet 处理大部分或者所有请求。令人遗憾的是，如果不修改源代码，将无法对 Servlet 进行注解。但值得庆幸的是，Servlet 3 中有一种比较容易的方法，能使一个 Servlet 变成一个 MultipartConfig Servlet，即给部署描述符（web.xml）中的 Servlet 声明赋值。以下代码与用@MultipartConfig 给 DispatcherServlet 进行注解的效果一样。

```
<servlet>
    <servlet-name>springmvc</servlet-name>
    <servlet-class>
        org.springframework.web.servlet.DispatcherServlet
    </servlet-class>
    <init-param>
        <param-name>contextConfigLocation</param-name>
        <param-value>
            /WEB-INF/config/springmvc-config.xml
        </param-value>
    </init-param>
    <multipart-config>
        <max-file-size>20848820</max-file-size>
        <max-request-size>418018841</max-request-size>
        <file-size-threshold>1048576</file-size-threshold>
    </multipart-config>
</servlet>
```

此外，还需要在 Spring MVC 配件文件中使用一个不同的多部分解析器，像下面这样：

```
<bean id="multipartResolver"
        class="org.springframework.web.multipart.support.
➥ StandardServletMultipartResolver">
</bean>
```

upload2 应用程序展示了如何在 Servlet 3 及其更高版本的容器中处理文件上传问题。这是从 upload1 改写过来的，因此，domain 和 controller 类都非常相似。唯一的区别在于，现在的 web.xml 文件中包含了一个 multipart-config 元素。清单 11.6 展示了 upload2 的 web.xml 文件。

清单 11.6　upload2 的 web.xml 文件

```
<?xml version="1.0" encoding="UTF-8"?>
<web-app version="3.0"
    xmlns="http://java.sun.com/xml/ns/javaee"
    xmlns:xsi="http://www.w3.org/2001/XMLSchema-instance"
    xsi:schemaLocation="http://java.sun.com/xml/ns/javaee
➥ http://java.sun.com/xml/ns/javaee/web-app_3_0.xsd">

    <servlet>
        <servlet-name>springmvc</servlet-name>
        <servlet-class>
            org.springframework.web.servlet.DispatcherServlet
        </servlet-class>
        <init-param>
            <param-name>contextConfigLocation</param-name>
```

```
            <param-value>
                /WEB-INF/config/springmvc-config.xml
            </param-value>
        </init-param>
        <load-on-startup>1</load-on-startup>
        <multipart-config>
            <max-file-size>20848820</max-file-size>
            <max-request-size>418018841</max-request-size>
            <file-size-threshold>1048576</file-size-threshold>
        </multipart-config>
    </servlet>

    <servlet-mapping>
        <servlet-name>springmvc</servlet-name>
        <url-pattern>/</url-pattern>
    </servlet-mapping>
</web-app>
```

清单 11.7 展示了 upload2 的 Spring MVC 配置文件。

清单 11.7　upload2 的 Spring MVC 配置文件

```
<?xml version="1.0" encoding="UTF-8"?>
<beans xmlns="http://www.springframework.org/schema/beans"
    xmlns:xsi="http://www.w3.org/2001/XMLSchema-instance"
    xmlns:p="http://www.springframework.org/schema/p"
    xmlns:mvc="http://www.springframework.org/schema/mvc"
    xmlns:context="http://www.springframework.org/schema/context"
    xsi:schemaLocation="
        http://www.springframework.org/schema/beans
        http://www.springframework.org/schema/beans/spring-beans.xsd
        http://www.springframework.org/schema/mvc
        http://www.springframework.org/schema/mvc/spring-mvc.xsd
        http://www.springframework.org/schema/context
        http://www.springframework.org/schema/context/spring-
➥ context.xsd">

    <context:component-scan base-package="controller" />
    <mvc:annotation-driven />

    <mvc:resources mapping="/css/**" location="/css/" />
    <mvc:resources mapping="/*.html" location="/" />
    <mvc:resources mapping="/image/**" location="/image/" />
    <mvc:resources mapping="/file/**" location="/file/" />

    <bean id="viewResolver"
        class="org.springframework.web.servlet.view.
```

```
➥ InternalResourceViewResolver">
        <property name="prefix" value="/WEB-INF/jsp/" />
        <property name="suffix" value=".jsp" />
    </bean>

    <bean id="multipartResolver"
        class="org.springframework.web.multipart.support.
➥ StandardServletMultipartResolver">
    </bean>
</beans>
```

如果要对这个应用程序进行测试，请在浏览器中访问以下 URL：

```
http://localhost:8080/upload2/input-product
```

11.10　客户端上传

虽然 Servlet 3 中的文件上传特性使文件上传变得十分容易，只需在服务器端编程即可，但这对提升用户体验却毫无帮助。单独一个 HTML 表单并不能显示进度条，或者显示已经成功上传的文件数量。开发人员采用了各种不同的技术来改善用户界面，例如，单独用一个浏览器线程对服务器发出请求，以便报告上传进度，或者利用像 Java applets、Adobe Flash、Microsoft Silverlight 这样的第三方技术。

这些第三方技术可以工作，但都在一定程度上存在限制。今天 Java applets 和 Silverlight 几乎死了，Chrome 不再允许 applet 和 Silverlight，Microsoft 取代 Internet Explorer 的新浏览器 Edge 根本就不支持插件。

你仍然可以使用 Flash，因为 Chrome 仍然可以运行它，Edge 已经集成了它。然而，现在越来越多的人选择拥抱 HTML5。

HTML 5 在其 DOM 中添加了一个 File API。它允许访问本地文件。与 Java 小程序、Adobe Flash、Microsoft Silverlight 相比，HTML 5 似乎是针对客户端文件上传局限性的最佳解决方案。令人遗憾的是，在编写本书时，IE 9 尚未完全支持这个 API，但可以利用最新版的 Firefox、Chrome 和 Opera 浏览器来测试下面的例子。

为了证明 HTML 5 的威力，upload2 中的 html5.jsp 页面采用了 JavaScript 和 HTML 5 File API 来提供报告上传进度的进度条。upload2 应用程序中也复制了一份 MultipleUploadsServlet 类，用于在服务器中保存已上传的文件。但是，JavaScript 不在本书讨论范围之内，因此这里只做简单的说明。

简言之，我们关注的是 HTML 5 input 元素的 change 事件，当 input 元素的值发生改变时，就会触发它。本书还关注 HTML 5 在 XMLHttpRequest 对象中添加的 progress 事件。XMLHttpRequest 自然是 AJAX 的骨架。当异步使用 XMLHttpRequest 对象上传文件时，就会持续地触发 progress 事件，直到上传进度完成或取消，或者直到上传进度因为出错而中断。通过监听 progress 事件，可以轻松地监测文件上传操作的进度。

upload2 中的 Html5FileUploadController 类能够将已经上传的文件保存到应用程序目录的 file 目录下。清单 11.8 中的 UploadedFile 类展示了一个简单的 domain 类，它只包含一个属性。

清单 11.8 UploadedFile 的 domain 类

```
package domain;
import java.io.Serializable;
import org.springframework.web.multipart.MultipartFile;

public class UploadedFile implements Serializable {
    private static final long serialVersionUID = 1L;
    private MultipartFile multipartFile;
    public MultipartFile getMultipartFile() {
        return multipartFile;
    }
    public void setMultipartFile(MultipartFile multipartFile) {
        this.multipartFile = multipartFile;
    }
}
```

Html5FileUploadController 类如清单 11.9 所示。

清单 11.9 Html5FileUploadController 类

```
package controller;
import java.io.File;
import java.io.IOException;
import javax.servlet.http.HttpServletRequest;
import org.apache.commons.logging.Log;
import org.apache.commons.logging.LogFactory;
import org.springframework.stereotype.Controller;
import org.springframework.ui.Model;
import org.springframework.validation.BindingResult;
import org.springframework.web.bind.annotation.ModelAttribute;
import org.springframework.web.bind.annotation.RequestMapping;
import org.springframework.web.multipart.MultipartFile;
```

```
import domain.UploadedFile;

@Controller
public class Html5FileUploadController {

    private static final Log logger = LogFactory
            .getLog(Html5FileUploadController.class);

    @RequestMapping(value = "/html5")
    public String inputProduct() {
        return "Html5";
    }

    @RequestMapping(value = "/file_upload")
    public void saveFile(HttpServletRequest servletRequest,
            @ModelAttribute UploadedFile uploadedFile,
            BindingResult bindingResult, Model model) {

        MultipartFile multipartFile =
                uploadedFile.getMultipartFile();
        String fileName = multipartFile.getOriginalFilename();
        try {
            File file = new File(servletRequest.getServletContext()
                    .getRealPath("/file"), fileName);
            multipartFile.transferTo(file);
        } catch (IOException e) {
            e.printStackTrace();
        }
    }
}
```

Html5FileUploadController 中的 saveFile 方法将已经上传的文件保存到应用程序目录中的 file 目录下。

清单 11.10 所示的 html5.jsp 页面中包含的 JavaScript 代码允许用户选择多个文件，并且一键单击即可全部上传。这些文件本身将同时上传。

清单 11.10　html5.jsp 页面

```
<!DOCTYPE HTML>
<html>
<head>
<script>
    var totalFileLength, totalUploaded, fileCount, filesUploaded;
```

```
function debug(s) {
    var debug = document.getElementById('debug');
    if (debug) {
        debug.innerHTML = debug.innerHTML + '<br/>' + s;
    }
}
function onUploadComplete(e) {
    totalUploaded += document.getElementById('files').
            files[filesUploaded].size;
    filesUploaded++;
    debug('complete ' + filesUploaded + " of " + fileCount);
    debug('totalUploaded: ' + totalUploaded);
    if (filesUploaded < fileCount) {
        uploadNext();
    } else {
        var bar = document.getElementById('bar');
        bar.style.width = '100%';
        bar.innerHTML = '100% complete';
        alert('Finished uploading file(s)');
    }
}
function onFileSelect(e) {
    var files = e.target.files; // FileList object
    var output = [];
    fileCount = files.length;
    totalFileLength = 0;
    for (var i=0; i<fileCount; i++) {
        var file = files[i];
        output.push(file.name, ' (',
              file.size, ' bytes, ',
              file.lastModifiedDate.toLocaleDateString(), ')'
        );
        output.push('<br/>');
        debug('add ' + file.size);
        totalFileLength += file.size;
    }
    document.getElementById('selectedFiles').innerHTML =
        output.join('');
    debug('totalFileLength:' + totalFileLength);
}

function onUploadProgress(e) {
    if (e.lengthComputable) {
        var percentComplete = parseInt(
                (e.loaded + totalUploaded) * 100
```

```
                    / totalFileLength);
            var bar = document.getElementById('bar');
            bar.style.width = percentComplete + '%';
            bar.innerHTML = percentComplete + ' % complete';
        } else {
            debug('unable to compute');
        }
    }

    function onUploadFailed(e) {
        alert("Error uploading file");
    }

    function uploadNext() {
        var xhr = new XMLHttpRequest();
        var fd = new FormData();
        var file = document.getElementById('files').
                files[filesUploaded];
        fd.append("multipartFile", file);
        xhr.upload.addEventListener(
                "progress", onUploadProgress, false);
        xhr.addEventListener("load", onUploadComplete, false);
        xhr.addEventListener("error", onUploadFailed, false);
        xhr.open("POST", "file_upload");
        debug('uploading ' + file.name);
        xhr.send(fd);
    }
    function startUpload() {
        totalUploaded = filesUploaded = 0;
        uploadNext();
    }
    window.onload = function() {
        document.getElementById('files').addEventListener(
                'change', onFileSelect, false);
        document.getElementById('uploadButton').
                addEventListener('click', startUpload, false);
    }
</script>
</head>
<body>
<h1>Multiple file uploads with progress bar</h1>
<div id='progressBar' style='height:20px;border:2px solid green'>
    <div id='bar'
            style='height:100%;background:#33dd33;width:0%'>
    </div>
```

```
</div>
<form>
    <input type="file" id="files" multiple/>
    <br/>
    <output id="selectedFiles"></output>
    <input id="uploadButton" type="button" value="Upload"/>
</form>
<div id='debug'
    style='height:100px;border:2px solid green;overflow:auto'>
</div>
</body>
</html>
```

html5.jsp 页面的用户界面中主要包含了一个名为 progressBar 的 div 元素、一个表单和另一个名为 debug 的 div 元素。也许你已经猜到了，progressBar div 用于展示上传进度，debug 用于调试信息。表单中有一个类型为 file 的 input 元素和一个按钮。

这个表单中有两点需要注意。第一，是标识为 files 的 input 元素，它有一个 multiple 属性，用于支持多文件选择。第二，这个按钮不是一个提交按钮。因此，单击它并不会提交表单。事实上，脚本是利用 XMLHttpRequest 对象来完成上传的。

下面来看 JavaScript 代码。我们假定读者已经具备一定的脚本语言知识。

执行脚本时，它做的第一件事就是为这 4 个变量分配空间：

```
var totalFileLength, totalUploaded, fileCount, filesUploaded;
```

totalFileLength 变量保存要上传的文件总长度。totalUploaded 是指目前已经上传的字节数。fileCount 中包含了要上传的文件数量。filesUploaded 表示已经上传的文件数量。

随后，当窗口完全下载后，便调用赋予 window.onload 的函数。

```
window.onload = function() {
    document.getElementById('files').addEventListener(
            'change', onFileSelect, false);
    document.getElementById('uploadButton').
            addEventListener('click', startUpload, false);
}
```

这段代码将 files input 元素的 change 事件映射到 onFileSelect 函数，将按钮的 click 事件映射到 startUpload。

每当用户从本地目录中修改了不同的文件时，都会触发 change 事件。与该事件相关的事件处理器只是在一个 output 元素中输出已选中的文件的名称和容量。下面是一个事件处理器的例子：

```
function onFileSelect(e) {
    var files = e.target.files; // FileList object
    var output = [];
    fileCount = files.length;
    totalFileLength = 0;
    for (var i=0; i<fileCount; i++) {
        var file = files[i];
        output.push(file.name, ' (',
              file.size, ' bytes, ',
              file.lastModifiedDate.toLocaleDateString(), ')'
        );
        output.push('<br/>');
        debug('add ' + file.size);
        totalFileLength += file.size;
    }
    document.getElementById('selectedFiles').innerHTML =
        output.join('');
    debug('totalFileLength:' + totalFileLength);
}
```

当用户单击 Upload 按钮时，就会调用 startUpload 函数，并随之调用 uploadNext 函数。uploadNext 上传已选文件列表中的下一个文件。它首先创建一个 XMLHttpRequest 对象和一个 FormData 对象，并将接下来要上传的文件添加到它的后面。

```
var xhr = new XMLHttpRequest();
var fd = new FormData();
var file = document.getElementById('files').
        files[filesUploaded];
fd.append("multipartFile", file);
```

随后，uploadNext 函数将 XMLHttpRequest 对象的 progress 事件添加到 onUploadProgress，并将 load 事件和 error 事件分别添加到 onUploadComplete 和 onUploadFailed。

```
xhr.upload.addEventListener(
        "progress", onUploadProgress, false);
xhr.addEventListener("load", onUploadComplete, false);
xhr.addEventListener("error", onUploadFailed, false);
```

接下来，打开一个服务器连接，并发出 FormData。

```
xhr.open("POST", "file_upload");
debug('uploading ' + file.name);
xhr.send(fd);
```

在上传期间，会重复地调用 onUploadProgress 函数，让它有机会更新进度条。更新包括计算已经上传的总字节数比率，计算已选择文件的字节数，拓宽 progressBar div 元素里面的 div 元素。

```
function onUploadProgress(e) {
    if (e.lengthComputable) {
        var percentComplete = parseInt(
                (e.loaded + totalUploaded) * 100
                / totalFileLength);
        var bar = document.getElementById('bar');
        bar.style.width = percentComplete + '%';
        bar.innerHTML = percentComplete + ' % complete';
    } else {
        debug('unable to compute');
    }
}
```

上传完成时，调用 onUploadComplete 函数。这个事件处理器会增加 totalUploaded，即已经完成上传的文件容量，并添加 filesUploaded 值。随后，它会查看已经选中的所有文件是否都已经上传完毕。如果是，则会显示一条消息，告诉用户文件上传已经成功完成。如果不是，则再次调用 uploadNext。为了便于阅读，将 onUploadComplete 函数重新复制到这里。

```
function onUploadComplete(e) {
    totalUploaded += document.getElementById('files').
            files[filesUploaded].size;
    filesUploaded++;
    debug('complete ' + filesUploaded + " of " + fileCount);
    debug('totalUploaded: ' + totalUploaded);
    if (filesUploaded < fileCount) {
        uploadNext();
    } else {
        var bar = document.getElementById('bar');
        bar.style.width = '100%';
        bar.innerHTML = '100% complete';
        alert('Finished uploading file(s)');
    }
}
```

利用下面的 URL 可以对上述应用程序进行测试：

```
http://localhost:8080/upload2/html5
```

选择几个文件，并单击 Upload 按钮，将会看到一个进度条，以及上传文件的信息，屏幕截图如图 11.3 所示。

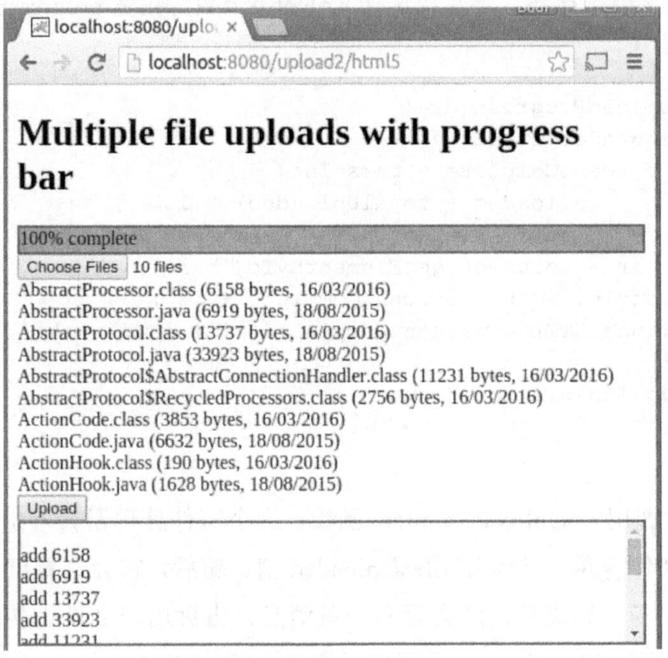

图 11.3　带进度条的文件上传

11.11　小结

本章介绍了如何在 Spring MVC 应用程序中处理文件上传。处理已上传的文件有两种方法，即利用 Commons FileUpload 组件，或者利用 Servlet 3 本地文件上传特性。本章提供的范例展示了如何使用这两种方法。

本章还介绍了如何利用 HTML 5 支持多文件上传，并利用 File API 提升客户端的用户体验。

第 12 章
下载文件

像图片或者 HTML 文件这样的静态资源，在浏览器中打开正确的 URL 即可下载。只要该资源是放在应用程序的目录下，或者放在应用程序目录的子目录下，而不是放在 WEB-INF 下，Servlet/JSP 容器就会将该资源发送到浏览器。然而，有时静态资源是保存在应用程序目录之外，或者是保存在某一个数据库中，或者有时需要控制它的访问权限，防止其他网站交叉引用它。如果出现以上任意一种情况，都必须通过编程来发送资源。

简言之，通过编程进行的文件下载，使你可以有选择地将文件发送到浏览器。本章将介绍如何通过编程把资源发送到浏览器，并举两个范例。

12.1 文件下载概览

为了将像文件这样的资源发送到浏览器，需要在控制器中完成以下工作：

（1）对请求处理方法使用 void 返回类型，并在方法中添加 HttpServletResponse 参数。

（2）将响应的内容类型设为文件的内容类型。Content-Type 标题在某个实体的 body 中定义数据的类型，并包含媒体类型和子类型标识符。欲了解标准的内容类型，请访问 http://www.iana.org/assignments/media-types。如果不清楚内容类型，并且希望浏览器始终显示 Sava As（另存为）对话框，则将它设为 APPLICATION/OCTET-STREAM。这个值是不区分大小写的。

（3）添加一个名为 Content-Disposition 的 HTTP 响应标题，并赋值 attachment; filename=fileName，这里的 fileName 是默认文件名，应该出现在 File Download（文件下载）对话框中。它通常与文件同名，但是也并非一定如此。

例如，以下代码将一个文件发送到浏览器：

```
FileInputStream fis = new FileInputStream(file);
BufferedInputStream bis = new BufferedInputStream(fis);
byte[] bytes = new byte[bis.available()];
response.setContentType(contentType);
OutputStream os = response.getOutputStream();
bis.read(bytes);
os.write(bytes);
```

为了编程将一个文件发送到浏览器，首先要读取该文件作为 FileInputStream，并将内容加载到一个字节数组。随后，获取 HttpServletResponse 的 OutputStream，并调用其 write 方法传入字节数组。

将文件发送到 HTTP 客户端的更好方法是使用 Java NIO 的 Files.copy()方法：

```
Path file = Paths.get(...);
Files.copy(file, response.getOutputStream());
```

代码更短，运行速度更快。

12.2　范例 1：隐藏资源

download 应用程序展示了如何向浏览器发送文件。在这个应用程序中，由 ResourceController 类处理用户登录，并将一个 secret.pdf 文件发送给浏览器。secret.pdf 文件放在 WEB-INF/data 目录下，因此不可能直接访问。只有得到授权的用户，才能看到它。如果用户没有登录，应用程序就会跳转到登录页面。

清单 12.1 中的 ResourceController 类提供了一个控制器，负责发送 secret.pdf 文件。只有当用户的 HttpSession 中包含一个 loggedIn 属性时，表示该用户已经成功登录，才允许该用户访问。

清单 12.1　ResourceController 类

```
package controller;
import java.io.IOException;
import java.nio.file.Files;
import java.nio.file.Path;
import java.nio.file.Paths;
import javax.servlet.http.HttpServletRequest;
import javax.servlet.http.HttpServletResponse;
import javax.servlet.http.HttpSession;
import org.apache.commons.logging.Log;
```

```
import org.apache.commons.logging.LogFactory;
import org.springframework.stereotype.Controller;
import org.springframework.ui.Model;
import org.springframework.web.bind.annotation.ModelAttribute;
import org.springframework.web.bind.annotation.RequestMapping;
import domain.Login;

@Controller
public class ResourceController {

    private static final Log logger =
        LogFactory.getLog(ResourceController.class);

    @RequestMapping(value="/login")
    public String login(@ModelAttribute Login login, HttpSession
            session, Model model) {
        model.addAttribute("login", new Login());
        if ("paul".equals(login.getUserName()) &&
                "secret".equals(login.getPassword())) {
            session.setAttribute("loggedIn", Boolean.TRUE);
            return "Main";
        } else {
            return "LoginForm";
        }
    }

    @RequestMapping(value="/download-resource")
    public String downloadResource(HttpSession session,
            HttpServletRequest request, HttpServletResponse response,
            Model model) {
        if (session == null ||
                session.getAttribute("loggedIn") == null) {
            model.addAttribute("login", new Login());
            return "LoginForm";
        }
        String dataDirectory = request.
                getServletContext().getRealPath("/WEB-INF/data");
        Path file = Paths.get(dataDirectory, "secret.pdf");
        if (Files.exists(file)) {
            response.setContentType("application/pdf");
            response.addHeader("Content-Disposition",
                    "attachment; filename=secret.pdf");

            try {
                Files.copy(file, response.getOutputStream());
            } catch (IOException ex) {
```

```
            }
        }
        return null;
    }
}
```

控制器中的第一个方法 login，将用户带到登录表单。

LoginForm.jsp 页面如清单 12.2 所示。

清单 12.2　LoginForm.jsp 页面

```
<%@ taglib prefix="form" uri="http://www.springframework.org/tags/form" %>
<%@ taglib prefix="c" uri="http://java.sun.com/jsp/jstl/core" %>
<!DOCTYPE HTML>
<html>
<head>
<title>Login</title>
<style type="text/css">@import url("<c:url
       value="/css/main.css"/>");</style>
</head>
<body>
<div id="global">
<form:form commandName="login" action="login" method="post">
    <fieldset>
        <legend>Login</legend>
        <p>
            <label for="userName">User Name: </label>
            <form:input id="userName" path="userName"
                    cssErrorClass="error"/>
        </p>
        <p>
            <label for="password">Password: </label>
            <form:password id="password" path="password"
                cssErrorClass="error"/>
        </p>
        <p id="buttons">
            <input id="reset" type="reset" tabindex="4">
            <input id="submit" type="submit" tabindex="5"
                value="Login">
        </p>
    </fieldset>
</form:form>
</div>
</body>
</html>
```

成功登录所用的用户名和密码必须在 login 方法中进行硬编码。例如，用户名必须为 paul，密码必须为 secret。如果用户成功登录，他就会被转到 Main.jsp 页面（清单 12.3）。Main.jsp 页面中包含了一个链接，用户可以单击它来下载文件。

清单 12.3 Main.jsp 页面

```
<%@ taglib uri="http://java.sun.com/jsp/jstl/core" prefix="c" %>
<!DOCTYPE HTML>
<html>
<head>
<title>Download Page</title>
<style type="text/css">@import url("<c:url
       value="/css/main.css"/>");</style>
</head>
<body>
<div id="global">
    <h4>Please click the link below.</h4>
    <p>
        <a href="resource_download">Download</a>
    </p>
</div>
</body>
</html>
```

ResourceController 类中的第二个方法 downloadResource，它通过验证 session 属性 loggedIn 是否存在，来核实用户是否已经成功登录。如果找到该属性，就会将文件发送给浏览器。如果没有找到，用户就会转到登录页面。注意，如果使用 Java 7 或其更高版本，则可以使用其新的 try-with-resources 特性，从而更加安全地处理资源。

通过调用以下 URL 中的 FileDownloadServlet，可以测试 download 应用程序。

```
http://localhost:8080/download/login
```

12.3 范例 2：防止交叉引用

心怀叵测的竞争对手有可能通过交叉引用“窃取”你的网站资产，例如，将你的资料公然放在他的网站上，好像那些东西原本就属于他的一样。如果通过编程控制，使得只有当 referer 标题中包含你的域名时才发出资源，就可以防止那种情况发生。当然，那些心意坚决的窃贼仍然有办法下载到你的东西，但是绝不会像以前那样不费吹灰之力就能得到。

download 应用程序利用清单 12.4 中的 ImageController 类，使得仅当 referer 标题不为 null

时，才将图片发送给浏览器。这样可以防止仅在浏览器中输入网址就能下载图片的情况发生。

清单 12.4　ImageController 类

```
package controller;
import java.io.IOException;
import java.nio.file.Files;
import java.nio.file.Path;
import java.nio.file.Paths;
import javax.servlet.http.HttpServletRequest;
import javax.servlet.http.HttpServletResponse;
import org.apache.commons.logging.Log;
import org.apache.commons.logging.LogFactory;
import org.springframework.stereotype.Controller;
import org.springframework.web.bind.annotation.PathVariable;
import org.springframework.web.bind.annotation.RequestHeader;
import org.springframework.web.bind.annotation.RequestMapping;
import org.springframework.web.bind.annotation.RequestMethod;

@Controller
public class ImageController {

    private static final Log logger =
            LogFactory.getLog(ImageController.class);

    @RequestMapping(value="/image_get/{id}", method = RequestMethod.GET)
    public void getImage(@PathVariable String id,
            HttpServletRequest request, HttpServletResponse response,
            @RequestHeader String referer) {
        if (referer != null) {
            String imageDirectory = request.getServletContext().
                    getRealPath("/WEB-INF/image");
            Path file = Paths.get(dataDirectory, id + ".jpg");
            if (Files.exists(file)) {
                response.setContentType("image/jpg");
                try {
                    Files.copy(file, response.getOutputStream());
                } catch (IOException ex) {
                    e.printStackTrace();
                }
            }
        }
    }
}
```

原则上，ImageController 类的作用与 ResourceController 无异。getImage 方法开头处的 if 语句，可以确保只有当 referer 标题不为 null 时，才发出图片。

利用清单 12.5 中的 images.html 文件，可以对这个应用程序进行测试。

清单 12.5　images.html 文件

```
<!DOCTYPE html>
<html>
<head>
    <title>Photo Gallery</title>
</head>
<body>
<img src="get-image/1"/>
<img src="get-image/2"/>
<img src="get-image/3"/>
<img src="get-image/4"/>
<img src="get-image/5"/>
<img src="get-image/6"/>
<img src="get-image/7"/>
<img src="get-image/8"/>
<img src="get-image/9"/>
<img src="get-image/10"/>
</body>
</html>
```

要想看到 ImageServlet 的效果，请在浏览器中打开以下网址：

```
http://localhost:8080/download/images.html
```

图 12.1 展示了使用 ImageServlet 后的效果。

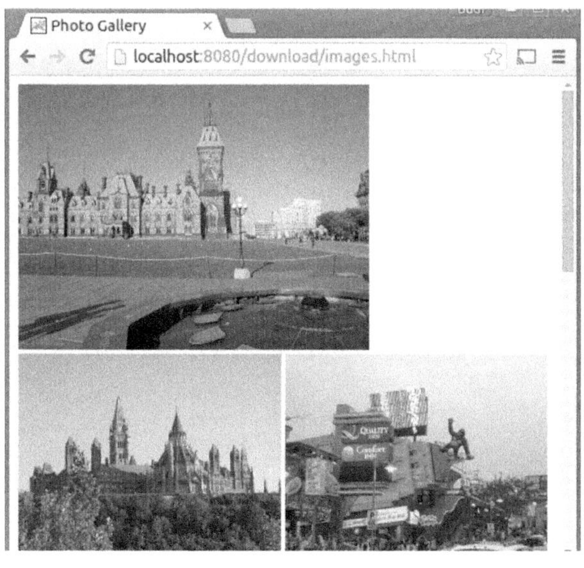

图 12.1　使用 ImageServlet 后的效果

12.4　小结

本章学习了如何在 Spring MVC 应用程序中通过编程来控制文件的下载，还学习了如何选择文件，以及如何将它发送给浏览器。

第 13 章 应用测试

测试在软件开发中的重要性不言而喻。测试的主要目的是尽早发现错误，最好是在代码开发的同时。逻辑上认为，错误发现的越早，修复的成本越低。如果你在编程时发现错误，可以立即更改代码；如果队友在你将代码发布到共享存储库之后发现了错误，则现在至少涉及两个人；想象一下，当软件发布后，客户发现错误后所需要的修复成本会更大。

在软件开发中有许多不同的测试。其中两个是单元测试和集成测试。你通常从单元测试开始测试类中的单个方法，然后进行集成测试，以测试不同的模块是否可以无缝协同工作。

本章中的示例使用 JUnit 测试框架以及 Spring-test 模块。Spring-test 模块中的 API 可用于单元测试和集成测试。你可以在 org.springframework.test 及其子包以及 org.springframework.mock.* 包中找到 Spring 测试相关的类型。本章附带的示例分为两个项目，称为单元测试和集成测试。第一个项目包含类和单元测试。第二个项目包含类和集成测试。

13.1 单元测试

单元测试的理想情况是为每个类创建一个测试类，并为类中的每个方法创建一个测试方法，像 getter 和 setter 方法这样的简单方法除外，它们直接从字段返回值或赋值给字段。

在测试用语中，被测试的类称为被测系统（SUT）。

单元测试旨在快速且多次运行。单元测试仅验证代码本身，而不涉及它的依赖，其任何依赖应该被帮助对象替代，这将在本章后面解释。涉及依赖性的测试通常在集成测试中完成，而不是单元测试。

乍一看，写单元测试看起来像不必要的额外工作。毕竟，你可以使用 main 方法从类本身内测试一个类。但是，你应该考虑单元测试的好处。首先，在单独的测试类中编写测试代码不会混淆你的类；第二，单元测试可以用于回归测试，在一些逻辑发生变化时，以确保一切仍然工作；单元测试的另一个主要优点是，它可以在持续集成设置中自动化测试。持续集成是指一种开发方法，当程序员将他们的代码提交到到共享库时，每次代码提交将触发一次自动构建并运行所有单元测试。持续集成可以尽早地检测问题。

在单元测试中，类使用 new 运算符实例化。不依赖 Spring 框架的依赖注入容器来创建 bean。

让我们来看看清单 13.1 中的类。

清单 13.1　被测试类

```
package com.example.util;
public class MyUtility {

    public int method1(int a, int b) { ... }
    public long method2(long a) { ... }

}
```

为了对这个类进行单元测试，创建一个如清单 13.2 所示的类。注意每个方法应该至少有一个测试方法。

清单 13.2　测试类

```
package com.example.util;
public class MyUtilityTest {

    public void testMethod1() {
        MyUtility utility = new MyUtility();
        int result = utility.method1(100, 200);
        // assert that result equals the expected value
    }

    public void testMethod2() {
        MyUtility utility = new MyUtility();
        long result = utility.method2(100L);
        // assert that result equals the expected value
    }
}
```

单元测试有些约定俗成，首先是将测试类命名为与带有 Test 的 SUT 相同的名称。因此，MyUtility 的测试类应命名为 MyUtilityTest；其次，测试类的包路径应与 SUT 相同，以允许前者访问后者的受保护和默认成员。但是，测试类应位于不同于测试的类的源文件夹下。

测试方法没有返回值。在测试方法中，你实例化要测试的类，调用要测试的方法并验证结果。为了使测试类更容易编写，你应该使用测试框架，例如 JUnit 或 TestNG。本章介绍用 JUnit 编写测试类的例子，JUnit 是 Java 的事实上的标准单元测试框架。

13.2　状态测试与行为测试

大多数时候，你只关心方法是否返回正确的结果。这称为状态测试。但是有时候，你还需要执行行为测试（也称为交互测试），以确保方法的行为正确。例如，你可能想要验证方法以正确的顺序调用其他方法。或者，你可能需要确保你的方法在操作期间调用另一个方法正好 n 次。

你将在本章后面学习如何执行状态测试和行为测试。

13.3　应用 JUnit

对于单元测试，我推荐使用 JUnit，你可以从 http://junit.org 下载它。点击下载并安装链接后，你将被重定向到下载页面。你需要下载 junit.jar 和 hamcrest-core.jar 文件。后者是 JUnit 的依赖项，目前版本为 4.12。

如果使用 Maven 或 STS，请将此元素添加到 pom.xml 文件以下载 JUnit 及其依赖关系。

```
<dependency>
    <groupId> junit </ groupId>
    <artifactId> junit </ artifactId>
    <version> 4.12 </ version>
    <scope>test</scope>
</ dependency>
```

JUnit 也广泛用于集成测试，你将在本章后面了解这一点。

你可以通过阅读 JUnit 网站上的文档，了解有关 JUnit 的更多信息以及如何编写 JUnit 测试。

13.3.1　开发一个单元测试

编写单元测试很容易。使用@org.junit.Test 简单地注解所有测试方法。此外，你可以通过使用@ org.junit.Before 注解来创建初始化方法。初始化方法在调用任何测试方法之前调用，你可以编写准备测试方法的代码，例如创建将由测试方法使用的对象。

你还可以通过使用@ org.junit.After 注解方法来创建清理方法。清理方法在测试类中的所有测试方法都被执行之后调用，并且可以用来释放测试期间使用的资源。

清单 13.3 显示了需要进行单元测试的 Calculator 类。

清单 13.3　Calculator 类

```
package com.example;
public class Calculator {
    public int add(int a, int b) {
        return a + b;
    }

    public int subtract(int a, int b) {
        return a - b;
    }
}
```

清单 13.4 显示了 Calculator 类的单元测试类。

清单 13.4　单元测试类

```
package com.example;
import org.junit.After;
import org.junit.Assert;
import org.junit.Before;
import org.junit.Test;

public class CalculatorTest {
    @Before
    public void init() {
    }

    @After
    public void cleanUp() {
    }

    @Test
    public void testAdd() {
        Calculator calculator = new Calculator();
        int result = calculator.add(5, 8);
        Assert.assertEquals(13, result);
    }

    @Test
    public void testSubtract() {
        Calculator calculator = new Calculator();
        int result = calculator.subtract(5, 8);
```

```
        Assert.assertEquals(-3, result);
    }
}
```

CalculatorTest 类有两个测试方法、一个初始化方法（用@Before 注解）和一个清除方法（用@After 注解）。org.junit.Assert 类提供用于声明结果的静态方法。例如，assertEquals 方法用来比较两个值。

13.3.2 运行一个单元测试

Eclipse 和 STS 知道一个类是否是一个 JUnit 测试类。要运行测试类，请右键单击包资源管理器中的测试类，然后选择运行方式> JUnit 测试。或者，使用编辑器打开测试类，然后按 Ctrl+F11 键。

测试完成后，如果 JUnit 视图尚未打开，Eclipse 或 STS 将打开它。如果单元测试成功完成，你将在 JUnit 视图中看到一个绿色条，如图 13.1 所示。

如果其中一个测试失败，你将看到一个红色条，如图 13.2 所示。

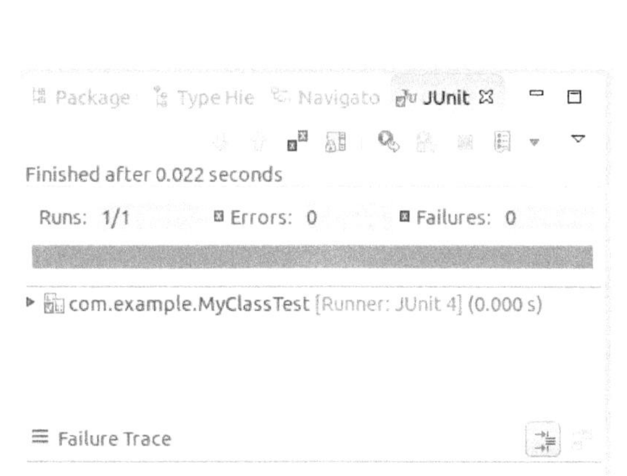

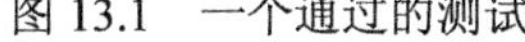
图 13.1　一个通过的测试

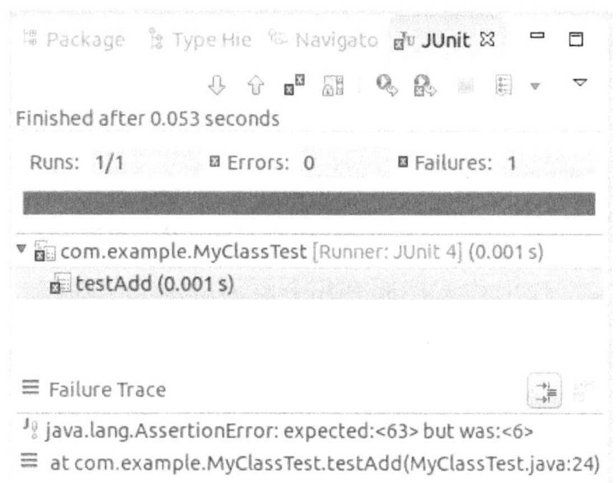

图 13.2　一个没通过的测试

13.3.3 通过测试套件来运行全部或多个单元测试

在有十几个类的一个小项目中，你将有十几个测试类。在一个更大的项目中，你会有更多测试类。在 Eclipse 或 STS 中，运行一个测试类很容易，但如何运行所有的测试类？

使用 JUunit 的解决方案非常简单。创建一个 Java 类并使用@RunWith（Suite.class）和@SuiteClasses()注解它。后者应该列出你想运行的所有类和其他套件测试。清单 13.5 显示了一个示例。类体可以留空。

清单 13.5 一个测试套件

```
package com.example;
import org.junit.runner.RunWith;
import org.junit.runners.Suite;
import org.junit.runners.Suite.SuiteClasses;

@RunWith(Suite.class)
@SuiteClasses({ MyTest1.class, MyTest2.class })
public class MyTestSuite {
}
```

如果使用 CI 服务器，则可以将服务器设置为每次提交新代码或修改的代码时自动运行单元测试。

13.4 应用测试挡板（Test Doubles）

被测系统（SUT）很少孤立存在。通常为了测试一个类，你需要依赖。在测试中，你的 SUT 所需要的依赖称为协作者。

协作者经常被称为测试挡板[①]的其他对象取代。测试挡板是 Gerard Meszaros 在他的书《xUnit Test Patterns: Refaltoning your code》中创造的一个术语。他对该术语的解释可以在这个网页中找到：http://xunitpatterns.com/Test%20Double.html

使用测试挡板有几个原因。

- 在编写测试类时，真正的依赖还没有准备好。
- 一些依赖项，例如 HttpServletRequest 和 HttpServletResponse 对象，是从 servlet 容器获取的，而自己创建这些对象将会非常耗时。
- 一些依赖关系启动和初始化速度较慢。例如，DAO 对象访问数据库导致单元测试执行很慢。

测试挡板在单元测试中广泛使用，也用于集成测试。当前有许多用于创建测试挡板的框架。Spring 也有自己的类来创建测试挡板。

模拟框架可用于创建测试挡板和验证代码行为。这里有一些流行的框架。

- Mockito。
- EasyMock。

① 译者注：Test Doubles 又翻译为测试替身，本书翻译为测试挡板。

- jMock。

除了上面的库，Spring 还附带了创建模拟对象的类。在下一节中，我们将使用 Spring 类来测试 Spring MVC 控制器。在本节中，我们将学习使用 Mockito。

Mockito 无法直接下载，所以你必须使用 Maven。但是，本章的示例项目包括 Mockito 发布包（一个 mockito.jar 文件）及其依赖（一个 objenesis.jar 文件）。

要使用 Maven 下载 Mockito，请将以下依赖关系添加到 pom.xml 文件中。

```
<dependency>
    <groupId>org.mockito</groupId>
    <artifactId>mockito-core</artifactId>
    <version>2.0.43-beta</version>
    <type>pom</type>
</dependency>
```

在开始写测试挡板之前，你应该先学习理论知识。如下是测试挡板的 5 种类型。

- dummy；
- stub；
- spy；
- fake；
- mock。

这些类型中的每一种将在下面的小节中解释。

13.4.1　dummy

dummy 是最基本的测试挡板类型。一个 dummy 是一个协作者的实现，它不做任何事情，并且不改变 SUT 的行为。它通常用于使 SUT 可以实例化。dummy 只在开发的早期阶段使用。

例如，考虑清单 13.6 中的 ProductServiceImpl 类。这个类依赖于传递给构造函数的 ProductDAO。

清单 13.6　ProductServiceImpl 类

```
package com.example.service;
import com.example.dao.ProductDAO;

public class ProductServiceImpl implements ProductService {

    private ProductDAO productDAO;
```

```
    public ProductServiceImpl(ProductDAO productDAOArg) {
        if (productDAOArg == null) {
            throw new NullPointerException("ProductDAO cannot be null.");
        }
        this.productDAO = productDAOArg;
    }

    @Override
    public BigDecimal calculateDiscount() {
        return productDAO.calculateDiscount();
    }
    @Override
    public boolean isOnSale(int productId) {
        return productDAO.isOnSale(productId);
    }
}
```

ProductServiceImpl 类需要一个非空的 ProductDAO 对象来实例化。同时，要测试的方法不使用 ProductDAO。因此，可以创建一个如清单 13.7 所示的 dummy 对象，只需让 ProductServiceImpl 可实例化。

清单 13.7 ProductDAODummy 类

```
package com.example.dummy;
import java.math.BigDecimal;
import com.example.dao.ProductDAO;

public class ProductDAODummy implements ProductDAO {
    public BigDecimal calculateDiscount() {
        return null;
    }
    public boolean isOnSale(int productId) {
        return false;
    };
}
```

在 dummy 类中的方法实现什么也不做，它的返回值也不重要，因为这些方法从未使用过。

清单 13.8 显示了一个可以运行的测试类

清单 13.8 ProductDAOTest 类

```
package com.example.dummy;
import static org.junit.Assert.assertNotNull;
import org.junit.Test;
```

```
import com.example.dao.ProductDAO;
import com.example.service.ProductService;
import com.example.service.ProductServiceImpl;

public class ProductServiceImplTest {

    @Test
    public void testCalculateDiscount() {
        ProductDAO productDAO = new ProductDAODummy();
        ProductService productService =
                new ProductServiceImpl(productDAO);
        assertNotNull(productService);
    }
}
```

13.4.2 stub

像 dummy 一样，stub 也是依赖接口的实现。和 dummy 不同的是，stub 中的方法返回硬编码值，并且这些方法被实际调用。

清单 13.9 显示了一个 stub，可用于测试清单 13.6 中的 ProductServiceImpl 类。

清单 13.9 ProductDAOStub 类

```
package com.example.stub;
import java.math.BigDecimal;
import com.example.dao.ProductDAO;

public class ProductDAOStub implements ProductDAO {
    public BigDecimal calculateDiscount() {
        return new BigDecimal(14);
    }
    public boolean isOnSale(int productId) {
        return false;
    };
}
```

13.4.3 spy

spy 是一个略微智能一些的 stub，因为 spy 可以保留状态。考虑下面的汽车租赁应用程序，其中包含一个 GarageService 接口和一个 GarageServiceImpl 类，分别在清单 13.10 和清单 13.11 中。

清单 13.10 GarageService 接口

```
package com.example.service;
import com.example.Car;
```

```
public interface GarageService {
    Car rent();
}
```

清单 13.11　GarageServiceImpl 类

```
package com.example.service;
import com.example.Car;
import com.example.dao.GarageDAO;

public class GarageServiceImpl implements GarageService {
    private GarageDAO garageDAO;
    public GarageServiceImpl(GarageDAO garageDAOArg) {
        this.garageDAO = garageDAOArg;
    }
    public Car rent() {
        return garageDAO.rent();
    }
}
```

GarageService 接口只有一个方法：rent。GarageServiceImpl 类是 GarageService 的一个实现，并且依赖一个 GarageDAO。GarageServiceImpl 中的 rent 方法调用 GarageDAO 中的 rent 方法。GarageDAO 的实现应该返回一个 car，如果还有一辆汽车在车库；或者返回 null，如果没有更多的汽车。

由于 GarageDAO 的真正实现还没有完成，清单 13.12 中的 GarageDAOSpy 类被用作测试挡板。它是一个 spy，因为它的方法返回一个硬编码值，并且它通过一个 carCount 变量来保存车库里的车数。

清单 13.12　GarageDAOSpy 类

```
package com.example.spy;
import com.example.Car;
import com.example.dao.GarageDAO;

public class GarageDAOSpy implements GarageDAO {
    private int carCount = 3;

    @Override
    public Car rent() {
        if (carCount == 0) {
            return null;
        } else {
            carCount--;
```

```
            return new Car();
        }
    }
}
```

清单 13.13 显示了使用 GarageDAOSpy 测试 GarageServiceImpl 类的一个测试类。

清单 13.13 GarageServiceImplTest 类

```
package com.example.spy;
import com.example.Car;
import com.example.dao.GarageDAO;
import com.example.service.GarageService;
import com.example.service.GarageServiceImpl;
import org.junit.Test;
import static org.junit.Assert.*;

public class GarageServiceImplTest {

    @Test
    public void testRentCar() {
        GarageDAO garageDAO = new GarageDAOSpy();
        GarageService garageService = new GarageServiceImpl(garageDAO);
        Car car1 = garageService.rent();
        Car car2 = garageService.rent();
        Car car3 = garageService.rent();
        Car car4 = garageService.rent();

        assertNotNull(car1);
        assertNotNull(car2);
        assertNotNull(car3);
        assertNull(car4);
    }
}
```

由于在车库只有 3 辆车，spy 只能返回 3 辆车，当第四次调用其 rent 方法时，返回 null。

13.4.4 fake

fake 的行为就像一个真正的合作者，但不适合生产，因为它走“捷径”。内存存储是一个 fake 的完美示例，因为它的行为像一个 DAO，但不会将其状态保存到硬盘驱动器。

例如，分别考虑清单 13.14 和清单 13.15 中的 Member 和 MemberServiceImpl 类。Member 类包含成员的标识符和名称。MemberServiceImpl 类可以将成员添加到商店并检索所有存储的成员。

清单 13.14　Member 类

```
package com.example.model;
public class Member {
    private int id;
    private String name;
    public Member(int idArg, String nameArg) {
        this.id = idArg;
        this.name = nameArg;
    }
    public int getId() {
        return id;
    }
    public void setId(int idArg) {
        this.id = idArg;
    }
    public String getName() {
        return name;
    }
    public void setName(String nameArg) {
        this.name = nameArg;
    }
}
```

清单 13.15　MemberServiceImpl 类

```
package com.example.service;
import java.util.List;
import com.example.dao.MemberDAO;
import com.example.model.Member;

public class MemberServiceImpl implements MemberService {
    private MemberDAO memberDAO;
    public void setMemberDAO(MemberDAO memberDAOArg) {
        this.memberDAO = memberDAOArg;
    }

    @Override
    public void add(Member member) {
        memberDAO.add(member);
    }
    @Override
    public List<Member> getMembers() {
        return memberDAO.getMembers();
    }
}
```

MemberServiceImpl 依赖于 MemberDAO。但是，由于没有可用的 MemberDAO 实现，你可以创建一个 MemberDAO 的 fake 实现，以便可以立即测试 MemberServiceImpl。清单 13.16 显示了这样一个 fake。它将成员存储在 ArrayList 中，而不是持久化存储。因此，不能在生产中使用它，但是对于单元测试是足够的。

清单 13.16 MemberDAOFake 类

```
package com.example.fake;
import java.util.ArrayList;
import java.util.List;
import com.example.dao.MemberDAO;
import com.example.model.Member;

public class MemberDAOFake implements MemberDAO {
    private List<Member> members = new ArrayList<>();

    @Override
    public void add(Member member) {
        members.add(member);
    }

    @Override
    public List<Member> getMembers() {
        return members;
    }
}
```

清单 13.17 显示了一个测试类，它使用 MemberDAOFake 作为 MemberDAO 的测试挡板类来测试 MemberServiceImpl。

清单 13.17 MemberServiceImplTest 类

```
package com.example.service;
import org.junit.Assert;
import org.junit.Test;
import com.example.dao.MemberDAO;
import com.example.fake.MemberDAOFake;
import com.example.model.Member;

public class MemberServiceImplTest {

    @Test
    public void testAddMember() {
        MemberDAO memberDAO = new MemberDAOFake();
        memberDAO.add(new Member(1, "John Diet"));
```

```
        memberDAO.add(new Member(2, "Jane Biteman"));
        Assert.assertEquals(2, memberDAO.getMembers().size());
    }
}
```

13.4.5　mock

mock 在理念上不同于其他测试挡板。使用 dummy、stub、spy 和 fake 来进行状态测试，即验证方法的输出。而使用 mock 来执行行为（交互）测试，以确保某个方法真正被调用，或者验证一个方法在执行另一个方法期间被调用了一定的次数。

例如，考虑清单 13.18 中的 MathUtil 类。

清单 13.18　MathUtil 类

```
package com.example;
public class MathUtil {
    private MathHelper mathHelper;
    public MathUtil(MathHelper mathHelper) {
        this.mathHelper = mathHelper;
    }
    public MathUtil() {

    }

    public int multiply(int a, int b) {
        int result = 0;
        for (int i = 1; i <= a; i++) {
            result = mathHelper.add(result, b);
        }
        return result;
    }
}
```

MathUtil 有一个方法 multiply。它非常直接，使用多个 add 方法。换句话说，3 × 8 计算为 8 + 8 + 8。MathUtil 类甚至不知道如何执行 add。为此，它依赖于 MathHelper 对象，其类在列表 13.19 中结出。

清单 13.19　MathHelper 类

```
package com.example;
public class MathHelper {
    public int add(int a, int b) {
```

```
        return a + b;
    }
}
```

测试所关心的不是 multiply 方法的结果，而是找出方法是否如预期一样运行。因此，在计算 3 × 8 时，它应该调用 MathHelper.add() 3 次。清单 13.20 显示了一个使用 MathHelper 模拟的测试类。Mockito 是一个流行的模拟框架，用于创建模拟对象。我将在本章稍后介绍关于 Mockito 的更多知识。在这一节中，我只是想告诉你这个概念。

清单 13.20 MathUtilTest 类

```
package com.example;
import static org.mockito.Mockito.mock;
import static org.mockito.Mockito.times;
import static org.mockito.Mockito.verify;
import static org.mockito.Mockito.when;
import org.junit.Test;

public class MathUtilTest {

    @Test
    public void testMultiply() {
        MathHelper mathHelper = mock(MathHelper.class);
        for (int i = 0; i < 10; i++) {
            when(mathHelper.add(i * 8, 8)).thenReturn(i * 8 + 8);
        }
        MathUtil mathUtil = new MathUtil(mathHelper);
        mathUtil.multiply(3, 8);
        verify(mathHelper, times(1)).add(0, 8);
        verify(mathHelper, times(1)).add(8, 8);
        verify(mathHelper, times(1)).add(16, 8);
    }
}
```

使用 Mockito 创建 mock 对象非常简单，只需调用 org.mockito.Mockito 类的静态 mock 方法即可。下面展示如何创建 MathHelper mock 对象。

```
MathHelper mathHelper = mock(MathHelper.class);
```

接下来，你需要使用 when 方法准备 mock 对象。基本上，你告诉它，给定使用这组参数的方法调用，mock 对象必须返回这个值。例如，这条语句是说如果调用 mathHelper.add(10, 20)，返回值必须是 10 + 20：

```
when(mathHelper.add(10, 20)).thenReturn(10 + 20);
```

对于此测试，你准备具有十组参数的 mock 对象（但不是所有的参数组都会被使用）。

```
for (int i = 0; i < 10; i++) {
    when(mathHelper.add(i * 8, 8)).thenReturn(i * 8 + 8);
}
```

然后创建要测试的对象并调用其参数。

```
MathUtil mathUtil = new MathUtil(mathHelper);
mathUtil.multiply(3, 8);
```

接下来的 3 条语句是行为测试。为此，调用 verify 方法：

```
verify(mathHelper, times(1)).add(0, 8);
verify(mathHelper, times(1)).add(8, 8);
verify(mathHelper, times(1)).add(16, 8);
```

第一条语句验证 mathHelper.add（0，8）被调用了一次。第二条语句验证 mathHelper.add（8,8）被调用一次，第三条语句验证 mathHelper.add（16，8）也被调用一次。

13.5 对 Spring MVC Controller 单元测试

你已经学习了如何在 Spring MVC 应用程序中测试各个类。但 Controller 有点不同，因为它们通常与 Servlet API 对象（如 HttpServletRequest、HttpServletResponse、HttpSession 等）交互。在许多情况下，你将需要模拟这些对象以正确测试控制器。

像 Mockito 或 EasyMock 这样的框架是可以模拟任何 Java 对象的通用模拟框架。你必须自己配置生成的对象（使用一系列 when 语句）。Spring-Test 模拟对象是专门为使用 Spring 而构建的，并且与真实对象更接近，更容易使用。

以下小节讨论了一些更重要的单元测试控制器类型。

13.5.1 MockHttpServletRequest 和 MockHttpServletResponse

当调用控制器时，你可能需要传递 HttpServletRequest 和 HttpServletResponse。在生产环境中，两个对象都由 servlet 容器本身提供。在测试环境中，你可以使用 org.springframework.mock.web 包中的 Spring MockHttpServletRequest 和 MockHttpServletResponse 类。

这两个类很容易使用。你可以通过调用其无参构造函数来创建实例：

```
MockHttpServletRequest request = new MockHttpServletRequest();
MockHttpServletResponse response = new MockHttpServletResponse();
```

MockHttpServletRequest 类实现了 javax.servlet.http.HttpServletRequest，并允许你将实例配置将看起来像一个真正的 HttpServletRequest。它提供了方法来设置 HttpServletRequest 中的所有属性以及获取其属性的值。表 13.1 显示了它的一些方法。

表 13.1 MockHttpServletRequest 的主要方法

方法	描述
addHeader	添加一个 HTTP 请求头
addParameter	添加一个请求参数
getAttribute	返回一个属性
getAttributeNames	返回包含了全部属性名的一个 Enumeration 对象
getContextPath	返回上下文路径
getCookies	返回全部的 cookies
setMethod	设置 HTTP 方法
setParameter	设置一个参数值
setQueryString	设置查询语句
setRequestURI	设置请求 URI

MockHttpServletResponse 实现了 javax.servlet.http.HttpServletResponse，并提供了配置实例的其他方法。表 13.2 显示了其一些主要的方法。

表 13.2 MockHttpServletResponse 的主要方法

方法	描述
addCookie	添加一个 cookie
addHeader	添加一个 HTTP 响应头
getContentLength	返回内容长度
getWriter	返回 Writer
getOutputStream	返回 ServletOutputStream

考虑清单 13.21 的 VideoController 类。

清单 13.21 VideoController 类

```
package com.example.controller;
import javax.servlet.http.HttpServletRequest;
import javax.servlet.http.HttpServletResponse;
import org.springframework.stereotype.Controller;
```

```
import org.springframework.web.bind.annotation.RequestMapping;

@Controller
public class VideoController {
    @RequestMapping(value = "/mostViewed")
    public String getMostViewed(HttpServletRequest request,
            HttpServletResponse response) {
        Integer id = (Integer) request.getAttribute("id");
        if (id == null) {
            response.setStatus(500);
        } else if (id == 1) {
            request.setAttribute("viewed", 100);
        } else if (id == 2) {
            request.setAttribute("viewed", 200);
        }
        return "mostViewed";
    }
}
```

VideoController 类的 getMostViewed 方法中，若请求属性 id 存在且值为 1 或 2，则添加请求属性“viewed”。否则，不添加请求属性。

清单 13.22 中的 VideoControllerTest 类使用两个测试方法验证这一点。

清单 13.22　VideoControllerTest 类

```
package com.example.controller;
import org.junit.Test;
import static org.junit.Assert.*;
import org.springframework.mock.web.MockHttpServletRequest;
import org.springframework.mock.web.MockHttpServletResponse;

public class VideoControllerTest {
    @Test
    public void testGetMostViewed() {
        VideoController videoController = new VideoController();
        MockHttpServletRequest request = new MockHttpServletRequest();
        request.setRequestURI("/mostViewed");
        request.setAttribute("id", 1);
        MockHttpServletResponse response = new MockHttpServletResponse();

        videoController.getMostViewed(request, response);
        assertEquals(200, response.getStatus());
        assertEquals(100L, (int) request.getAttribute("viewed"));
    }
```

```
    @Test
    public void testGetMostViewedWithNoId() {
        VideoController videoController = new VideoController();
        MockHttpServletRequest request = new MockHttpServletRequest();
        request.setRequestURI("/mostViewed");
        MockHttpServletResponse response = new MockHttpServletResponse();

        videoController.getMostViewed(request, response);
        assertEquals(500, response.getStatus());
        assertNull(request.getAttribute("viewed"));
    }
}
```

testGetMostViewed 方法实例化 VideoController 类并创建两个 mock 对象，一个 MockHttpServletRequest 和一个 MockHttpServletResponse。它还设置请求 URI，并向 MockHttpServletRequest 添加属性“id”。

```
VideoController videoController = new VideoController();
MockHttpServletRequest request = new MockHttpServletRequest();
request.setRequestURI("/mostViewed");
request.setAttribute("id", 1);
MockHttpServletResponse response = new MockHttpServletResponse();
```

然后调用 VideoController 的 getMostView 方法，传递 mock 对象，然后验证响应的状态码为 200，请求包含一个值为 100 的“viewed”属性。

```
videoController.getMostViewed(request, response);
assertEquals(200, response.getStatus());
assertEquals(100L, (int) request.getAttribute("viewed"));
```

VideoControllerTest 中的第二个方法类似方法一，但不会向 MockHttpServletRequest 对象添加“id”属性。因此，在调用控制器的方法时，它接收 HTTP 响应状态代码 500，并且在 MockHttpServletRequest 对象中没有“viewed”属性。

13.5.2 ModelAndViewAssert

ModelAndViewAssert 类是 org.springframework.web.servlet 包的一部分，是另一个有用的 Spring 类，用于测试从控制器的请求处理方法返回的 ModelAndView。回想一下第 4 章中介绍过，ModelAndView 是请求处理方法可以返回的类型之一，是包含有关请求处理方法的模型和视图的信息的一个 bean。

表 13.3 列举了 ModelAndViewAssert 的一些主要方法。

表 13.3　ModelAndViewAssert 的主要方法

方法	描述
assertViewName	检查 ModelAndView 的视图名称是否与预期名称匹配
assertModelAttributeAvailable	检查 ModelAndView 的模型是否包含具有预期模型名称的属性
assertModelAttributeValue	检查 ModelAndView 模型是否包含具有指定名称和值的属性
assertSortAndCompareList-ModelAttribute	对 ModelAndView 的列表排序，然后将其与预期列表进行比较

例如，考虑清单 13.23 中的 Book 类。这是一个简单的 bean 类，有 4 个属性：isbn、title、author 和 pubDate

清单 13.23　Book 类

```
package com.example.model;
import java.time.LocalDate;

public class Book {
    private String isbn;
    private String title;
    private String author;
    private LocalDate pubDate;

    public Book(String isbn, LocalDate pubDate) {
        this.isbn = isbn;
        this.pubDate = pubDate;
    }

    public Book(String isbn, String title, String author,
            LocalDate pubDate) {
        this.isbn = isbn;
        this.title = title;
        this.author = author;
        this.pubDate = pubDate;
    }

    // getters and setters not shown to save space

    @Override
    public boolean equals(Object otherBook) {
        return isbn.equals(((Book)otherBook).getIsbn());
    }
}
```

清单 13.24 中的 BookController 类是一个 Spring MVC 控制器，它包含一个请求处理方法

getLatestTitles。该方法接受 pubYear 路径变量，并返回一个 ModelAndView，如果 pubYear 的值为“2016”，它将包含书籍列表。

清单 13.24 BookController 类

```
package com.example.controller;
import java.time.LocalDate;
import java.util.Arrays;
import java.util.List;
import org.springframework.stereotype.Controller;
import org.springframework.web.bind.annotation.PathVariable;
import org.springframework.web.bind.annotation.RequestMapping;
import org.springframework.web.servlet.ModelAndView;
import com.example.model.Book;

@Controller
public class BookController {
    @RequestMapping(value = "/latest/{pubYear}")
    public ModelAndView getLatestTitles(
            @PathVariable String pubYear) {
        ModelAndView mav = new ModelAndView("Latest Titles");

        if ("2016".equals(pubYear)) {
            List<Book> list = Arrays.asList(
                    new Book("0001", "Spring MVC: A Tutorial",
                            "Paul Deck",
                            LocalDate.of(2016, 6, 1)),
                    new Book("0002", "Java Tutorial",
                            "Budi Kurniawan", LocalDate.of(2016, 11, 1)),
                    new Book("0003", "SQL", "Will Biteman",
                            LocalDate.of(2016, 12, 12)));
            mav.getModel().put("latest", list);
        }
        return mav;
    }
}
```

测试 BookController 的一种简单方式是使用 ModelAndViewAssert 中的静态方法，如清单 13.25 中的 BookControllerTest 类所示。

清单 13.25 BookControllerTest 类

```
package com.example.controller;
import static org.springframework.test.web.ModelAndViewAssert.*;
import java.time.LocalDate;
```

```
import java.util.Arrays;
import java.util.Comparator;
import java.util.List;
import org.junit.Test;
import org.springframework.web.servlet.ModelAndView;
import com.example.model.Book;

public class BookControllerTest {
    @Test
    public void test() {
        BookController bookController = new BookController();
        ModelAndView mav = bookController.getLatestTitles("2016");
        assertViewName(mav, "Latest Titles");
        assertModelAttributeAvailable(mav, "latest");
        List<Book> expectedList = Arrays.asList(
                new Book("0002", LocalDate.of(2016, 11, 1)),
                new Book("0001", LocalDate.of(2016, 6, 1)),
                new Book("0003", LocalDate.of(2016, 12, 12)));
        assertAndReturnModelAttributeOfType(mav, "latest",
                expectedList.getClass());
        Comparator<Book> pubDateComparator =
                (a, b) -> a.getPubDate().compareTo(b.getPubDate());
        assertSortAndCompareListModelAttribute(mav, "latest",
                expectedList, pubDateComparator);
    }
}
```

13.6　应用 Spring MVC Test 进行集成测试

集成测试用来测试不同的模块是否可以一起工作。它还确保两个模块之间数据的传递。使用 Spring 框架依赖注入容器，必须检查 bean 依赖注入。

若没有合适的工具，集成测试可能需要很多时间。想象一下，如果你正在建立一个网上商店，你必须使用网络浏览来测试购物车是否正确计算。每次更改代码时，你必须启动浏览器，登录系统，将几个项目添加到购物车，并检查总数是否正确。每次迭代可以轻易地花费 5 分钟！

好在，Spring 提供了一个用于集成测试的模块：Spring Test。

Spring 的 MockHttpServletRequest、MockHttpServletResponse 和 ModelAndViewAssert 类适用于对 Spring MVC 控制器进行单元测试，但它们缺少与集成测试相关的功能。例如，它们直接调用请求处理方法，无法测试请求映射和数据绑定。它们也不测试 bean 依赖注入，因为 SUT 类使用 new 运算符实例化。

对于集成测试，您需要一组不同的 Spring MVC 测试类型。以下小节讨论集成测试的 API 并提供一个示例。

13.6.1 API

作为 Spring 的一个模块，Spring MVC Test 提供了一些实用程序类，可以方便地在 Spring MVC 应用程序上执行集成测试。bean 是使用 Spring 依赖注入器创建的，并从 ApplicationContext 中获取，就像在一个真正的 Spring 应用程序中一样。

MockMvc 类位于 org.springframework.test.web.servlet 包下，是 Spring MVC Test 中的主类，用于帮助集成测试。此类允许你使用预定义的请求映射来调用请求处理方法。这里是一种常见的创建 MockMvc 实例的方法：

```
MockMvc mockMvc = MockMvcBuilders.webAppContextSetup(webAppContext).build();
```

这里，webAppContext 是 WebApplicationContext 实例的一个引用，WebApplicationContext 是第 2 章中讨论的 ApplicationContext 的子类，每个 Spring 开发人员都应该熟悉。要获取一个 WebApplicationContext，你必须在测试类中声明这一点。

```
@Autowired private WebApplicationContext webAppContext;
```

MockMvcBuilder 读取一个 Spring 配置文件或为测试类定义的文件。我将讨论如何指定测试类的配置文件，但首先我想讨论 MockMvc。

MockMvc 是一个非常简单的类。事实上，它只有一个方法：perform，用于通过 URI 间接调用 Spring MVC 控制器。

perform 方法具有以下签名：

```
ResultActions perform(RequestBuilder requestBuilder)
```

要测试请求处理方法，你需要创建一个 RequestBuilder。好在，MockMvcRequestBuilders 类提供了与 HTTP 方法具有相同名称的静态方法：get、post、head、put、patch、delete 等。要使用 HTTP GET 方法测试控制器，你可以调用 get 方法，要使用 POST 测试，则调用 post 方法。这些静态方法也很容易使用，你只需要传递一个字符串——控制器的请求处理方法的 URI。

例如，要调用名为 getEmployee 的请求处理方法，你将编写如下代码：

```
ResultActions resultActions = mockMvc.perform(get("/ getEmployee"));
```

当然，你必须导入 MockMvcRequestBuilders 的静态 get 方法。

要验证测试是否成功，你需要调用 ResultActions 的 andExpect 方法。andExpect 方法签名如下：

```
ResultAction andExpect(ResultMatcher matcher)
```

注意，andExpect 返回 ResultActions 的另一个实例，这意味着可以链式调用多个 AndExpect，你可以在稍后的示例中看到这一点。

MockMvcResultMatchers 类提供了静态方法来轻松创建 ResultMatcher。MockMvcResultMatchers 属于 org.springframework.test.web.servlet.result 包。表 13.4 显示了它的一些方法。

表 13.4　MockMvcResultMatchers 的主要方法

方法	返回类型	描述
cookie	CookieResultMatchers	返回一个 ResultMatchers，用来断言 cookie 值
header	HeaderResultMatchers	返回一个 ResultMatchers，用来断言 HTTP 响应头部
model	ModelResultMatchers	返回一个 ResultMatchers，用来断言请求处理的模型
status	StatusResultMatchers	返回一个 ResultMatchers，用来断言 HTTP 响应状态
view	ViewResultMatchers	返回一个 ResultMatchers，用来断言请求处理的视图

例如，要确保控制器方法的请求映射正确，可以使用状态方法：

```
mockMvc.perform(get("/ getBook")).andExpect(status().isOk());
```

isOK 方法断言响应状态代码是 200，可以看到 MockMvc 及其相关类使得集成测试控制器变的非常容易。

13.6.2　Spring MVC 测试类的框架

了解了 Spring MVC Test 中的一些重要的 API，现在来看下 Spring MVC 测试类的框架。

```
import org.junit.After;
import org.junit.Before;
import org.junit.Test;
import org.junit.runner.RunWith;
import org.springframework.beans.factory.annotation.Autowired;
import org.springframework.test.context.ContextConfiguration;
import org.springframework.test.context.junit4.SpringJUnit4ClassRunner;
import org.springframework.test.context.web.WebAppConfiguration;
import org.springframework.test.web.servlet.MockMvc;
import org.springframework.test.web.servlet.setup.MockMvcBuilders;
import org.springframework.web.context.WebApplicationContext;

@RunWith(SpringJUnit4ClassRunner.class)
@WebAppConfiguration
@ContextConfiguration("...")
```

```
public class ProductControllerTest {
    @Autowired
    private WebApplicationContext webAppContext;

    private MockMvc mockMvc;

    @Before
    public void setup() {
        mockMvc = MockMvcBuilders.webAppContextSetup(webAppContext).build();
    }

    @After
    public void cleanup(){
    }

    @Test
    public void test1() throws Exception {
        mockMvc.perform(...).andExpect(...);
    }

    @Test
    public void test2() throws Exception {
        mockMvc.perform(...).adnExpect(...);
    }

    ...

}
```

首先，看下你要导入的类型。在导入列表的顶部，是来自 JUnit 和 Spring MVC Test 的类型。像单元测试类一样，Spring MVC 测试类可以包括用@Before 和@After 注解的方法。两种注解类型都是 JUnit 的一部分。

接下去，测试框架开始与单元测试有所不同。首先是测试类运行器。你需要一个 SpringJUnit4ClassRunner.class 在@RunWith 注解内：

```
@RunWith(SpringJUnit4ClassRunner.class)
```

这个 runner 允许你使用 spring。

然后，你需要添加如下注解类型：

```
@WebAppConfiguration
@ContextConfiguration("...")
```

WebAppConfiguration 注解类型用于声明为集成测试加载的 ApplicationContext 应该是

WebApplicationContext 类型。ContextConfiguration 注解类型告诉测试运行器如何加载和配置 WebApplicationContext。

除以上，测试类中还需要两个对象：

```
private WebApplicationContext webAppContext;
private MockMvc mockMvc;
```

13.6.3　示例

以下示例展示如何对 Spring MVC 控制器开展集成测试。

清单 13.26 中的配置文件展示了将被扫描的包。这个文件是一个典型的 Spring MVC 配置文件，但减去了任何资源映射和视图解析器。但是，你可以使用实际的配置文件。

清单 13.26　test-config.xml 文件

```
<?xml version="1.0" encoding="UTF-8"?>
<beans xmlns="http://www.springframework.org/schema/beans"
   xmlns:xsi="http://www.w3.org/2001/XMLSchema-instance"
   xmlns:p="http://www.springframework.org/schema/p"
   xmlns:mvc="http://www.springframework.org/schema/mvc"
   xmlns:context="http://www.springframework.org/schema/context"
   xsi:schemaLocation="
      http://www.springframework.org/schema/beans
      http://www.springframework.org/schema/beans/spring-beans.xsd
      http://www.springframework.org/schema/mvc
      http://www.springframework.org/schema/mvc/spring-mvc.xsd
      http://www.springframework.org/schema/context
      http://www.springframework.org/schema/context/spring-context.xsd">
   <context:component-scan base-package="controller"/>
   <context:component-scan base-package="service"/>
   <mvc:annotation-driven/>
</beans>
```

此示例中的 SUT 是清单 13.27 中的 EmployeeController 类。类中只有一个请求处理方法，getHighestPaid，它映射到/highest-paid。

清单 13.27　EmployeeController 类

```
package controller;
import org.springframework.beans.factory.annotation.Autowired;
import org.springframework.stereotype.Controller;
import org.springframework.ui.Model;
import org.springframework.web.bind.annotation.PathVariable;
```

```
import org.springframework.web.bind.annotation.RequestMapping;
import service.EmployeeService;
import domain.Employee;

@Controller
public class EmployeeController {

    @Autowired
    EmployeeService employeeService;

    @RequestMapping(value="/highest-paid/{category}")
    public String getHighestPaid(@PathVariable int category, Model model)
    {
        Employee employee = employeeService.getHighestPaidEmployee( category);
        model.addAttribute("employee", employee);
        return "success";
    }
}
```

清单 13.28 展示了 EmployeeController 的测试类。

清单 13.28 EmployeeControllerTest 类

```
package com.example.controller;
import static org.springframework.test.web.servlet.request.
➥ MockMvcRequestBuilders.get;
import static org.springframework.test.web.servlet.result.
➥ MockMvcResultHandlers.print;
import static org.springframework.test.web.servlet.result.
➥ MockMvcResultMatchers.model;
import static org.springframework.test.web.servlet.result.
➥ MockMvcResultMatchers.status;
import org.junit.After;
import org.junit.Before;
import org.junit.Test;
import org.junit.runner.RunWith;
import org.springframework.beans.factory.annotation.Autowired;
import org.springframework.test.context.ContextConfiguration;
import org.springframework.test.context.junit4.SpringJUnit4ClassRunner;
import org.springframework.test.context.web.WebAppConfiguration;
import org.springframework.test.web.servlet.MockMvc;
import org.springframework.test.web.servlet.setup.MockMvcBuilders;
import org.springframework.web.context.WebApplicationContext;

@RunWith(SpringJUnit4ClassRunner.class)
@WebAppConfiguration
@ContextConfiguration("test-config.xml")
```

```
public class EmployeeControllerTest {
    @Autowired
    private WebApplicationContext webAppContext;

    private MockMvc mockMvc;

    @Before
    public void setup() {
        this.mockMvc = MockMvcBuilders.webAppContextSetup(webAppContext)
                .build();
    }

    @After
    public void cleanUp() {

    }

    @Test
    public void testGetHighestPaidEmployee() throws Exception {
        mockMvc.perform(get("/highest-paid/2"))
                .andExpect(status().isOk())
                .andExpect(model().attributeExists("employee"))
                .andDo(print());
    }
}
```

EmployeeControllerTest 类包含一个 setUp 方法，它创建一个 MockMvc 对象。testGetHighestPaidEmployee 方法执行测试，并期望响应状态代码为 200，并且模型具有 employee 属性。

测试方法还调用 andDo(print())在响应对象中打印各种值。如果你的测试成功通过，你应该会看到类似的结果。

```
MockHttpServletRequest:
     HTTP Method = GET
     Request URI = /highest-paid/2
      Parameters = {}
         Headers = {}
Handler:
            Type = controller.EmployeeController
          Method = public java.lang.String
      controller.EmployeeController.getHighestPaid(int,org.springframe-
      work.ui.Model)

Async:
    Async started = false
```

```
        Async result = null

Resolved Exception:
                Type = null

ModelAndView:
           View name = success
                View = null
           Attribute = employee
               value = Xiao Ming ($200000)
              errors = []

FlashMap:
          Attributes = null

MockHttpServletResponse:
              Status = 200
       Error message = null
             Headers = {}
        Content type = null
                Body = 
       Forwarded URL = success
      Redirected URL = null
             Cookies = []
```

13.7　修改集成测试中 Web 根路径

默认情况下，使用@WebAppConfiguration 注解的 Spring 集成测试类将使用相对于项目目录的/src/main/webapp 目录作为根目录。这是一个 Maven 标准布局。如果你不使用 STS，而是使用不依赖于 Maven 的 IDE（如 Eclipse 或 NetBeans），在某些罕见情况下，这可能会出现问题。例如，如果你需要使用 ServletContext.getRealPath()的值，你将获得相对于/src/main/webapp 的值，而不是相对于你的 Web 应用程序目录的值。

然而，你也不能使用/src/main/webapp/WEB-INF/classes 目录作为 Eclipse 中的输出目录，因为这意味着将其嵌套在源目录中，而在 Eclipse 是禁止这样做的。好在，可以将实际应用程序目录传递到@WebAppConfiguration 来修复此问题。例如，以下将让任何 WebApplicationContext 使用/ approotdir 作为应用程序目录。

```
@WebAppConfiguration("/approotdir")
```

清单 13.29 展示了一个 WebAppController 类，其有一个方法返回在线资源的真实路径。

清单 13.29 WebAppController 类

```
package controller;
import javax.servlet.ServletContext;
import org.springframework.beans.factory.annotation.Autowired;
import org.springframework.stereotype.Controller;
import org.springframework.ui.Model;
import org.springframework.web.bind.annotation.RequestMapping;

@Controller
public class WebAppController {
    // ServletContext cannot be used as a method parameter, inject
    // instead.
    @Autowired
    private ServletContext servletContext;

    @RequestMapping(value="/getWebAppDir")
    public String getWebAppDirectory(Model model) {
        model.addAttribute("webAppDir", servletContext.getRealPath("/"));
        return "success";
    }
}
```

清单 13.30 展示了一个用于集成测试 WebAppController 类的测试类。注意传递给@WebAppConfiguration 注解的值。

清单 13.30 WebAppControllerTest 类

```
package com.example.controller;
import static org.springframework.test.web.servlet.request.
➥ MockMvcRequestBuilders.get;
import static org.springframework.test.web.servlet.result.
➥ MockMvcResultHandlers.print;
import static org.springframework.test.web.servlet.result.
➥ MockMvcResultMatchers.model;
import static org.springframework.test.web.servlet.result.
➥ MockMvcResultMatchers.status;
import org.junit.After;
import org.junit.Before;
import org.junit.Test;
import org.junit.runner.RunWith;
import org.springframework.beans.factory.annotation.Autowired;
import org.springframework.test.context.ContextConfiguration;
import org.springframework.test.context.junit4.SpringJUnit4ClassRunner;
import org.springframework.test.context.web.WebAppConfiguration;
import org.springframework.test.web.servlet.MockMvc;
```

```
import org.springframework.test.web.servlet.setUp.MockMvcBuilders;
import org .springframework.web.context.WebApplicationContext;

@RunWith(SpringJUnit4ClassRunner.class)
@WebAppConfiguration("/webapp")
@ContextConfiguration("test-config.xml")
public class WebAppControllerTest {
    @Autowired
    private WebApplicationContext webAppContext;

    private MockMvc mockMvc;

    @Before
    public void setup() {
        this.mockMvc =
       MockMvcBuilders.webAppContextSetup(webAppContext).build();
    }

    @After
    public void cleanUp() {
    }

    @Test
    public void testWebAppDir() throws Exception {
        mockMvc.perform(get("/getWebAppDir"))
                .andExpect(status().isOk())
                .andExpect(model().attributeExists("webAppDir"))
                .andDo(print());
    }
}
```

13.8 小结

测试是软件开发中的一个重要步骤，你应该在开发周期中尽早地执行单元测试和集成测试这两种类型的测试。单元测试用于类的功能性验证。在单元测试中，所涉及依赖通常被测试挡板替换，其可以包括 dummy、stub、spy、fake 和 mock 对象。JUnit 是一个流行的用于单元测试的框架，并且通常与 mocking 框架（如 Mockito 或 EasyMock）结合使用。

集成测试用于确保同一应用程序中的不同模块可以一起工作，同时确保请求映射和数据绑定也可以工作。Spring Test MVC 是一个 Spring 模块，它提供了一组 API，可以轻松地对 Spring 应用程序执行集成测试。在本章中，你还学习了如何使用 Spring Test MVC。

附录 A Tomcat

Tomcat 是当今最流行的 Servlet/JSP 容器。它是免费、成熟、开源的。为了运行本书附带的范例应用程序，需要 Tomcat 7 或其更高版本，或者其他兼容的 Servlet/JSP 容器才行。附录 A 将介绍如何快速安装和配置 Tomcat。

A.1 下载和配置 Tomcat

首先，从 http://tomcat.apache.org 网站下载 Tomcat 的最新版本。选用 zip 或 gz 格式的最新二进制发行版本。Tomcat 8 需要用 JRE 7 或 JDK7 来运行。早期版本的 Tomcat 要求是用 JDK 来运行，因为需要一个 java 编译器来编译 JSP 页面。从 5.5 开始，Tomcat 捆绑了 Eclipse 的 java 编译器，因此不再需要 JDK 了。

下载了 zip 或者 gz 文件后，要进行解压，随后就能在安装目录下看到几个目录。

在 bin 目录中，可以看到启动和终止 Tomcat 的程序。webapps 目录很重要，因为可以在那里部署应用程序。此外，conf 目录中还包含了配置文件，包括 server.xml 和 tomcat-users.xml 文件。lib 目录也值得关注，因为其中包含了编译 Servlet 和定制标签所需的 Servlet 和 JSP API。

解压完 zip 或者 gz 文件后，应设置 JRE_HOME 为 JRE 安装目录（或设置 JAVA_HOME 环境变量为 JDK 安装目录）。设置方式很简单，在 bin 目录中创建一个 setenv.sh（Linux 或 Mac OSX 环境）或 setenv.bat（Windows 环境）。该文件包含一行代码，将会在 Tomcat 启动时执行。

如下是一个 serenv.sh 文件示例：

```
JRE_HOME=/opt/java/jdk1.7.0_79
```

setenv.bat 文件如下：

```
set JAVA_HOME=c:\Program Files\Java\jdk1.7.0_06
```

对于 Windows 用户，还可以下载对应的 Windows 安装版本，安装起来会容易一些。

A.2 启动和终止 Tomcat

下载并解压好 Tomcat 二进制版本文件后，就可以运行 startup.bat 文件（Windows）或 startup.sh 文件（UNIX/Linux/Mac OS）来启动 Tomcat。这两个文件都放在 Tomcat 安装目录的 bin 目录下。默认情况下，Tomcat 在端口 8080 运行，因此可以在浏览器中打开以下网址：

```
http://localhost:8080
```

终止 Tomcat 时，运行 bin 目录下的 shutdown.bat 文件（Windows）或者 shutdown.sh 文件（UNIX/Linux/Mac OS）。

A.3 定义上下文

要将 Servlet/JSP 应用程序部署到 Tomcat，需要显式或隐式定义一个 Tomcat 上下文。在 Tomcat 中，每一个 Tomcat 上下文都表示一个 Web 应用程序。

显式定义 Tomcat 上下文有几种方法，包括：

- 在 Tomcat 的 conf/Catalina/localhost 目录下创建一个 XML 文件。
- 在 Tomcat 的 conf/server.xml 文件中添加一个 Context 元素。

如果决定给每一个上下文都创建一个 XML 文件，那么这个文件名就很重要，因为上下文路径是从文件名衍生得到的。例如，把一个 commerce.xml 文件放在 conf/Catalina/localhost 目录下，那么应用程序的上下文路径就是 commerce，并且可以利用以下 URL 调用一个资源：

```
http://localhost:8080/commerce/resourceName
```

上下文文件中必须包含一个 Context 元素，作为它的根元素。这个元素大多没有子元素，它是该文件中唯一的元素。例如，下面就是一个示例上下文文件，其中只有一行代码：

```
<Context docBase="C:/apps/commerce" reloadable="true"/>
```

这里唯一必要的属性是 docBase，它用来定义应用程序的位置。reloadable 属性是可选的，但是如果存在，并且它的值设为 true，那么一旦应用程序中 Java 类文件或者其他资源有任何增加、减少或者更新，Tomcat 都会侦测到，并且一旦侦测到这类变化，Tomcat 就会重新加载应用程序。在部署期间，建议将 reloadable 值设为 True，在生产期间，则不建议这么做。

当把上下文文件添加到指定目录时，Tomcat 就会自动加载应用程序。当删除这个文件时，Tomcat 就会自动卸载应用程序。

定义上下文的另一种方法是在 conf/server.xml 文件中添加一个 Context 元素。为此，要先打开文件，并在 Host 元素下创建一个 Context 元素。与前一种方法不同的是，此处定义上下文需要给上下文路径定义 path 属性。下面举一个例子：

```
<Host name="localhost" appBase="webapps" unpackWARs="true"
      autoDeploy="true">

    <Context path="/commerce"
            docBase="C:/apps/commerce"
            reloadable="true"
    />
</Host>
```

一般来说，不建议通过 server.xml 来管理上下文，因为只有重启 Tomcat 后，更新才能生效。不过，如果有很多应用程序需要测试，你也许会觉得使用 server.xml 比较理想，因为可以在一个文件中同时管理所有的应用程序。

最后，通过将一个 war 文件或者整个应用程序复制到 Tomcat 的 webapps 目录下，还可以隐式地部署应用程序。

关于 Tomcat 上下文的更多信息，请通过以下网址查阅：

http://tomcat.apache.org/tomcat-8.0-doc/config/context.html

A.4　定义资源

定义一个 JNDI 资源，应用程序便可以在 Tomcat 上下文定义中使用。资源用 Context 元素目录下的 Resource 元素表示。

例如，为了添加一个打开 MySQL 数据库连接的 DataSource 资源，首先要添加下面这个 Resource 元素：

```
<Context [path="/appName"] docBase="...">
    <Resource name="jdbc/dataSourceName"
        auth="Container"
        type="javax.sql.DataSource"
```

```
        username="..."
        password="..."
        driverClassName="com.mysql.jdbc.Driver"
        url="..."
    />
</Context>
```

关于 Resource 元素的更多信息，请到以下网址查阅：

```
http://tomcat.apache.org/tomcat-8.0-doc/jndi-resources-howto.html
```

A.5 安装 TLS 证书

Tomcat 支持 TLS，并且用它确保机密数据的传输，如身份证号码和信用卡信息等。利用 KeyTool 程序生成一个 public/private 键对，同时选择一家可信任的授权机构，来创建和签发数字证书。

一旦收到证书，并将它导入到 keystore 后，下一步就是在服务器上安装证书了。如果使用的是 Tomcat，将 keystore 复制到复制服务器上的某个位置，并对 Tomcat 进行配置即可。随后，打开 conf/server.xml 文件，并在<service>下添加以下 Connector 元素。

```
<Connector port="443"
    minSpareThreads="5"
    maxSpareThreads="75"
    enableLookups="true"
    disableUploadTimeout="true"
    acceptCount="100"
    maxThreads="200"

    scheme="https"
    secure="true"
    SSLEnabled="true"
    keystoreFile="/path/to/keystore"
    keyAlias="example.com"
    keystorePass="01secret02%%%"
    clientAuth="false"
    sslProtocol="TLS"
/>
```

以上粗体字部分的代码与 SSL 有关。请确保未使用 SSLEnabled 属性，或确保将其设置为 false，以防止旧浏览器使用不再安全的 SSL。

附录 B

Spring Tool Suite 和 Maven

Spring 提供了自己的集成开发环境（IDE），称为 Spring Tool Suite（STS），它可能是构建 Spring 应用程序的最佳 IDE 了。STS 捆绑 Maven 作为其默认依赖项管理工具，因此你不需要单独安装 Maven。

本附录提供了有关如何使用 STS 和 Maven 的简短教程。

B.1 安装 STS

Spring Tool Suite 是一个基于 Eclipse 的集成开发环境（IDE），它非常优秀，可能是使用 Spring 框架的最佳 IDE。它同时也是一个一站式工具箱，包括开发 Spring MVC 应用程序所需的库和应用程序，例如，它附带 Maven 和 Pivotal tc Server 的开发版本（一个 Tomcat 的定制版）。 如果你还没有决定使用哪个 IDE，我建议你试试 STS。

要开始使用 STS，请首先从此网页下载：

```
https://spring.io/tools/sts
```

STS 采用 zip 文件分发。下载 zip 文件后，将其解压到你选择的目录。zip 文件包含一个 sts-bundle 目录，其下又有 3 个目录：

1．法律文件，包含各种工具的许可协议。

2．pivotal-tc-server-developer-x.y.z.RELEASE。包含 Pivotal tc 服务器的开发人员版本 x.y.z。

3．sts-x.y.z.RELEASE。包含 STS，其中 x.y.z 代表为 STS 的主版本和次要版本。

Windows 版本的启动程序为 STS.exe 文件，Linux 版本的启动程序为 STS 文件。双击 exe 文件或从命令行运行 STS 文件以启动 STS。第一次运行 STS，系统将提示你选择工作区。单击“确定”，你将看到 STS 主页面，如图 B.1 所示。

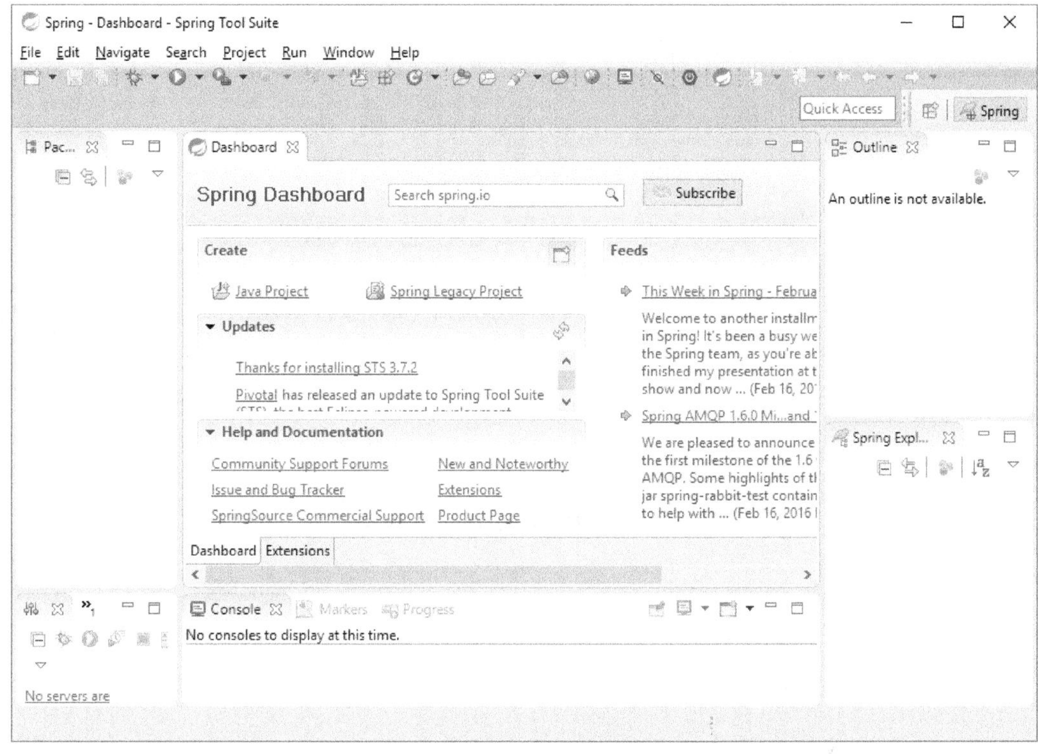

图 B.1 STS 欢迎界面

尽管已更改过图标，这个主窗口可能还是会让你想起 Eclipse。现在你可以随时创建自己的 Spring MVC 应用程序了。

B.2 创建一个 Spring MVC 应用

STS 严重依赖于 Maven，并允许你创建使用 Maven 管理依赖关系的应用程序。要创建 Spring MVC 应用程序，请按照下列步骤操作。

1. 点击 File > New > Maven Project，你会看到如图 B.2 所示的项目对话框。

2. 勾选“Create a simple project (skip archetype selection)”。

3. 点击 **Next** 按钮继续配置项目，将出现如图 B.3 所示的配置项。

4. 在“Group Id”输入框中输入包名，例如“com.example”；在“Artifact Id”输入框中输入项目名，例如“firstSpringMVC”。

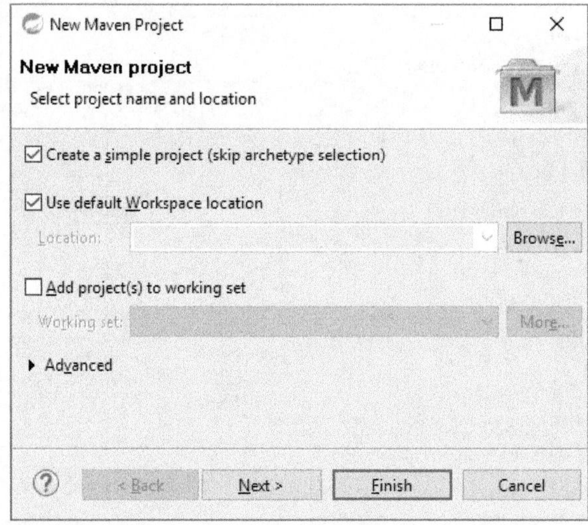

图 B.2　新 Maven 项目对话框

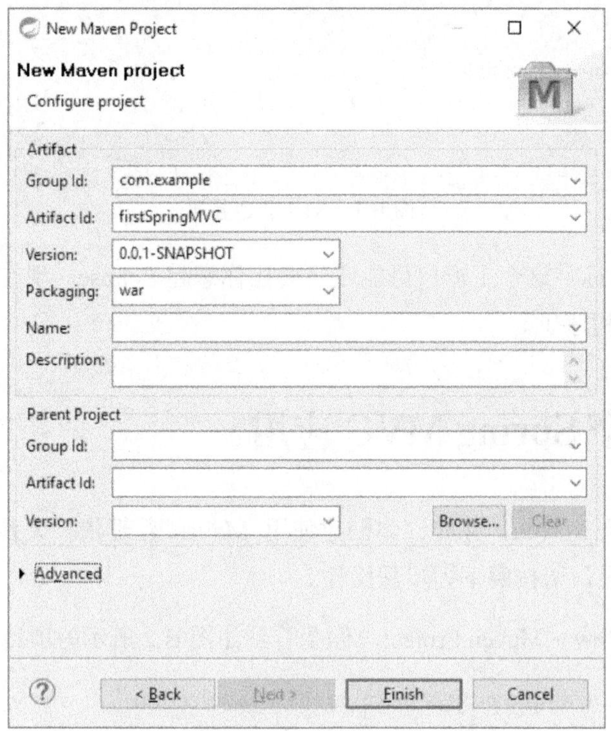

图 B.3　配置项目

5．在“Packaging”下拉框中选择“war”，告诉 STS 你将创建一个 Spring MVC 应用。war 是 servlet/JSP 应用的文件扩展名。

6．点击 Finish 按钮，你将在 Package Explorer 中看到你的项目（见图 B.4）。

7．STS 将会创建一个目录结构如图 B.4 所示的工程和一个 pom.xml 文件（maven 配置文件）。现在，你需要编辑 pom.xml，以便让 maven 来处理依赖。双击 pom.xml 文件，用默认编辑器打开文件。编辑器有多个选项卡，提供查看和编辑文件的不同方式。选项卡名称显示在对话框的底部。其中一个选项卡是“依赖项”选项卡，允许你管理项目所需的依赖项。单击依赖项选项卡，你将看到如图 B.5 所示的对话框。

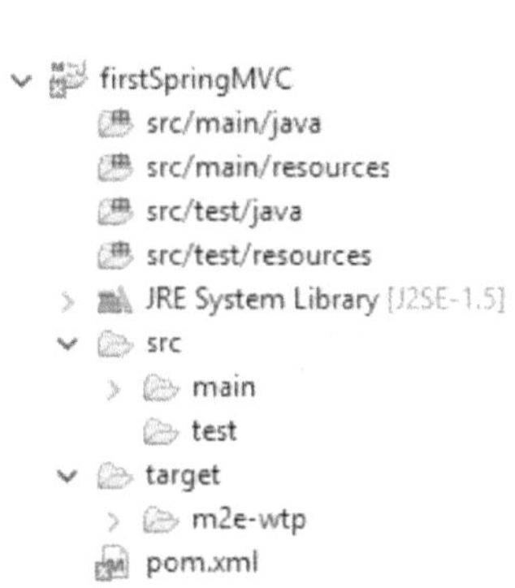

图 B.4　新创建的目录结构

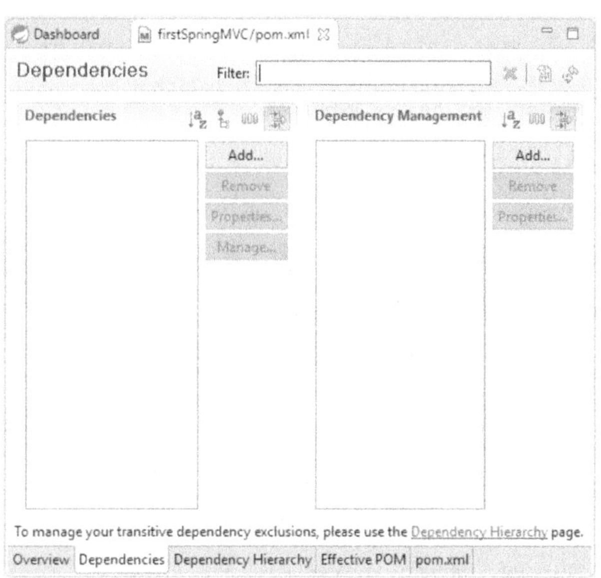

图 B.5　通过依赖选项卡添加依赖

8．对于初学者，通过 Dependencies 选项卡添加依赖是一种相对容易的方式，我将在这里展示如何做。典型的 Spring MVC 应用程序至少需要 3 个库：Servlet API、Spring MVC 库和 JSTL。要添加 Servlet API，请单击“Dependecies”窗格中的 Add 按钮（而不是“Denpendenly Management”窗格上的添加按钮），将打开“Select Dependency”对话框（见图 B.6）。

要输入的值如下。

```
Group Id: javax.servlet
Artifact Id: javax.servlet-api
Version: 3.1.0
Scope: provided
```

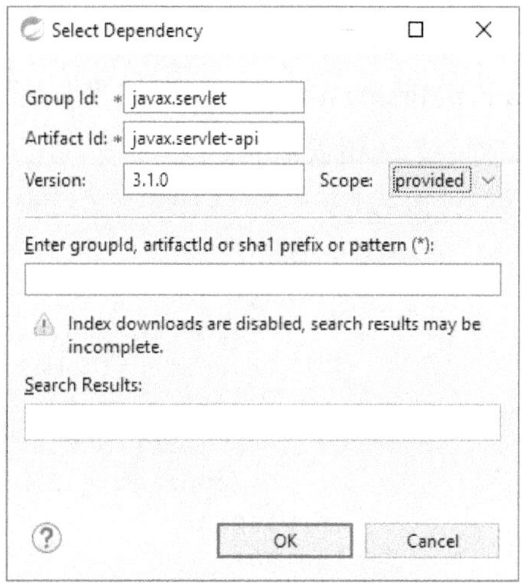

图 B.6　添加 Servlet API

provided 表示当应用程序分发时，该库将由容器提供，不需要包含在 war 文件中。

完成输入后，单击“OK”按钮，关闭该对话框，你将返回到“Dependencies”选项卡。

9．接下去，用同样的方式添加 Spring MVC 库，点击 Add 按钮，输入如下值：

```
Group Id: org.springframework
Artifact Id: spring-webmvc
Version: 4.2.4.RELEASE
Scope: compile
```

10．最后，添加 JSTL 库，点击 Add 按钮，输入如下值：

```
Group Id: javax.servlet
Artifact Id: jstl
Version: 1.2
Scope: runtime
```

11．按 Ctrl + S 或点击 File > Save 保存 pom.xml。这样，STS 将尝试通过下载依赖项来构建项目。你需要连接到互联网，除非你以前使用 Maven 下载过所需的库。

Maven 完成下载依赖项后，你将在 Package Explorer 中看到一个 Maven Dependencies 文件夹（见图 B.7）。

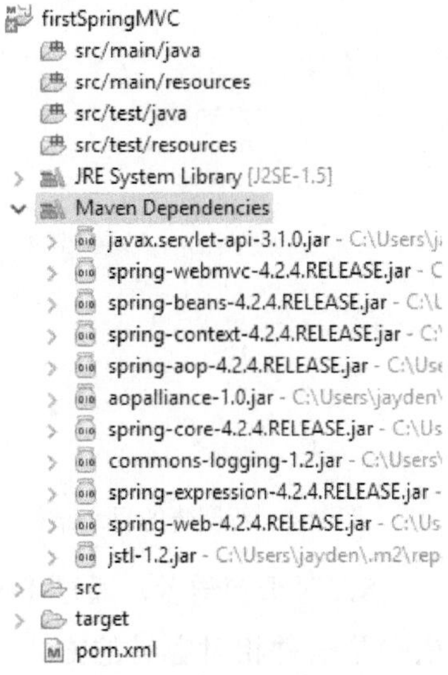

图 B.7 Maven 依赖项

pom.xml 文件仍然有错误，因为它找不到 web.xml。要纠正此问题，请双击 Package Explorer 中的 pom.xml 文件以重新打开编辑器，然后单击 pom.xml 选项卡。在 pom.xml 文件中查找<dependencies>元素，并在其上添加以下内容。注意，failOnMissingWebXml 元素用来打开或关闭由缺少 web.xml 文件引起的错误消息。

```
<build>
    <plugins>
        <plugin>
            <groupId>org.apache.maven.plugins</groupId>
            <artifactId>maven-war-plugin</artifactId>
            <configuration>
                <failOnMissingWebXml>false</failOnMissingWebXml>
            </configuration>
        </plugin>
    </plugins>
</build>
```

B.3 选择 Java 版本

您可以通过向 pom.xml 文件添加属性来为应用程序选择 Java 版本。请按照下列步骤操作。

1．点击 Overview 选项卡，如图 B.8 所示

2．点击属性下拉列表右侧的"Creat"按钮。将打开"Add Property"对话框，如图 B.9 所示。

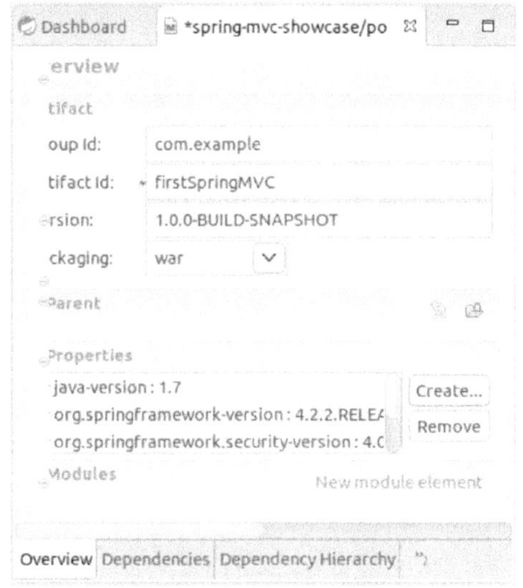

图 B.8 Overview 选项卡

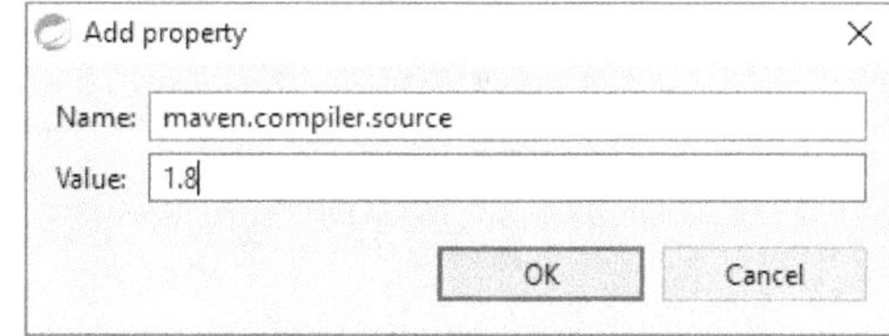

图 B.9 选择 Java 版本

3．在"Name"输入框中输入"maven.compiler.source"，在"Value"输入框中输入"1.8"。

4．保存 pom.xml 文件。

如果 STS 报错，那是因为它找不到 JRE 1.8，你可以通过执行如下步骤告诉它在哪里找到 Java。

1．点击 Window > Preferences。

2．在左窗格中选择 Java> Installed JRE，然后单击 Add 按钮。

3．选择 Standard VM，然后点击 Next。你将看到如图 B.10 所示的对话框。

4．点击"Directory"按钮并浏览到 JRE 主目录。

5．点击 Finish 按钮。您将在已安装的 JRE 列表中看到 JDK（见图 B.11）。

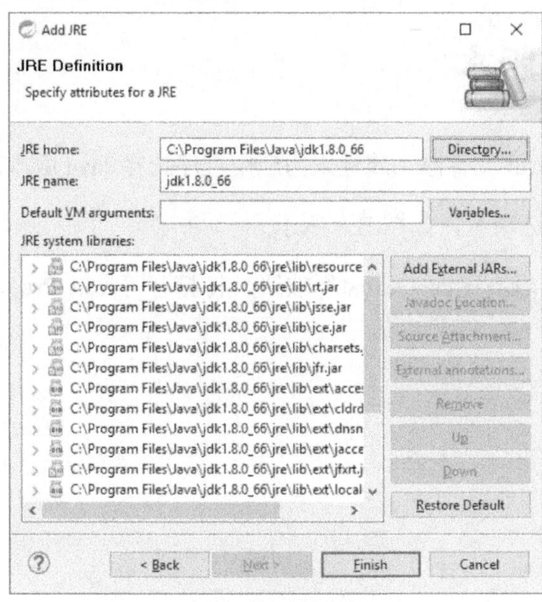

图 B.10　浏览 JRE 目录

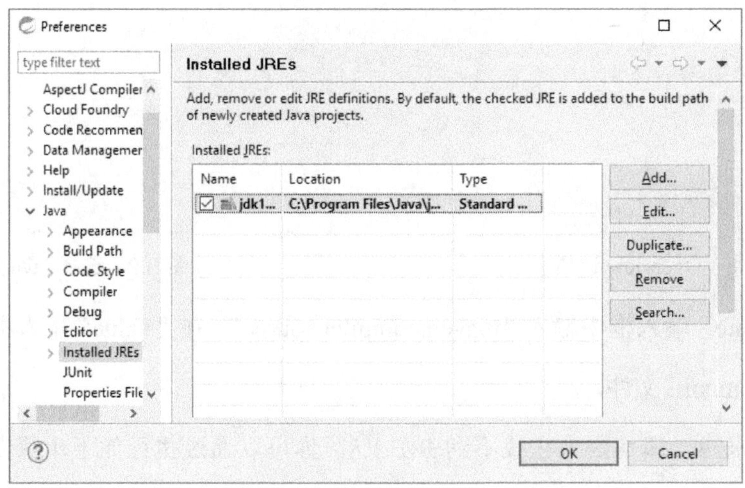

图 B.11　已安装的 JRE

6．点击 OK 按钮。

B.4　创建 index.html 文件

要完成示例应用程序，你需要创建一个 index.html 文件并将其保存在 src/main/webapp 下。

清单 B.1 显示了一个简单的 HTML 页面。

清单 B.1 index.html 文件

```
<!DOCTYPE html>
<html>
<head>
<title>First Spring MVC app</title></head>
<body>
Welcome
</body>
</html>
```

B.5 更新项目

在运行应用程序之前，你还要更新项目。右键单击 Package Explorer 中的项目图标，然后单击 Maven > Update Project。

之后，Package Explorer 中的项目将如图 B.12 所示。

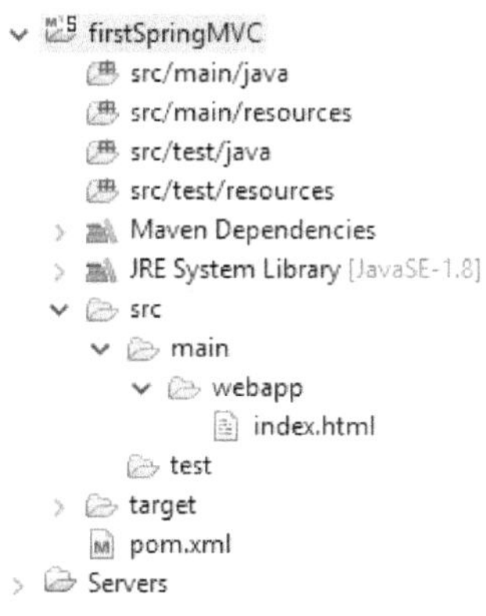

图 B.12 更新后的项目

B.6 运行应用

你需要一个 Tomcat 或其他的 servlet 容器来运行应用程序。如果你还没有安装 Tomcat，请现在安装。安装方式请参见附录 A。

请按照下列步骤运行应用：

1．右键单击项目并选择 Run As > Run on Server。你将看到 Run On Server 对话框，如图 B.13 所示。

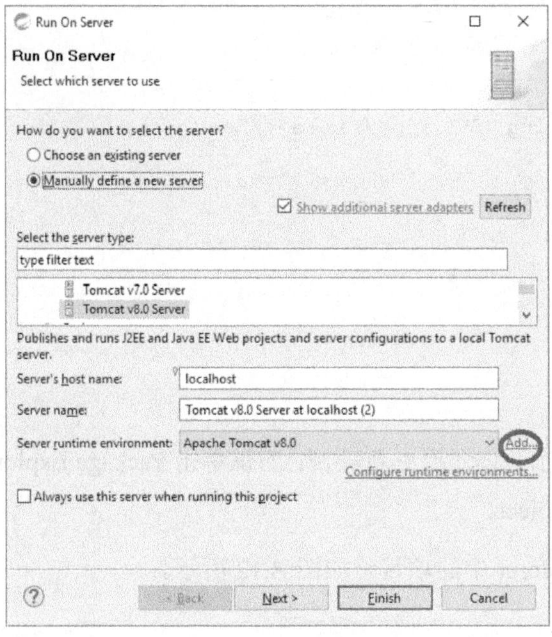

图 B.13　选择 Tomcat

2．选择服务器类型，例如 Apache 下的 Tomcat v8.0。如果没有看到 Tomcat 实例，请单击“Add”链接（图 B.13 中圈出），然后浏览到 Tomcat 目录。

3．在“Server Host Name”输入项中输入主机名。

4．单击 Finish 按钮。

STS 将启动 Tomcat 并运行你的应用程序。

要管理 Tomcat 及其上部署的所有应用程序，请打开 STS 中的服务器视图（见图 B.14）。

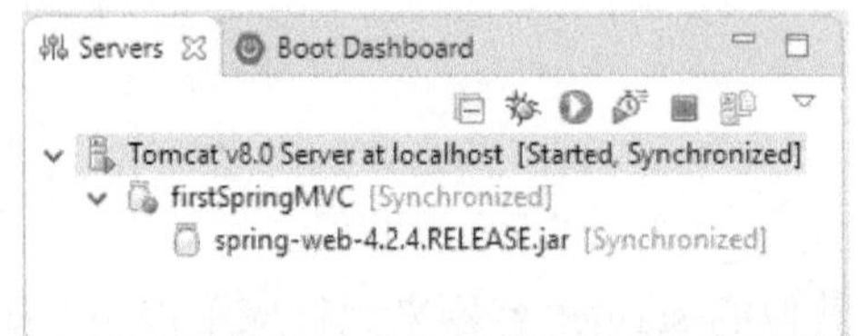

图 B.14　管理 Tomcat 实例和应用的服务器视图

图 B.15 显示了 STS 中的 Web 浏览器，其中显示了应用的默认页面。

图 B.15 你的第一个 Web 应用

现在，你可以按照每章中的教程来添加动态内容了。

附录 C Servlet

Servlet 是开发 Servlet 的主要技术。掌握 Servlet API 是成为一名强大的 Java Web 开发者的基本条件。你必须熟悉 Servlet API 中定义的 70 多种类型。“70”这个数字听起来很多，但是如果一次只学一种类型，其实并不困难。

本附录介绍了 Servlet API，并教你如何编写 Servlet。所有示例代码都可以在 servletapi1、servletapi2 和 servletapi3 中找到。

C.1 Servlet API 概览

Servlet API 有以下 4 个 Java 包：

- javax.servlet，其中包含定义 Servlet 和 Servlet 容器之间契约的类和接口。
- javax.servlet.http，其中包含定义 HTTP Servlet 和 Servlet 容器之间契约的类和接口。
- javax.servlet.annotation，其中包含用于 Servlet、filter、listener 的注解。它还为被注解元件定义元数据。
- javax.servlet.descriptor，其中包含提供程序化登录 Web 应用程序配置信息的类型。

本附录主要关注 javax.servlet 和 javax.servlet.http 的成员。

图 C.1 中展示了 javax.servlet 中的主要类型。

Servlet 技术的核心是 Servlet，它是所有 Servlet 类必须直接或间接实现的一个接口。在编写实现 Servlet 的 Servlet 类时，直接实现它。在扩展实现这个接口的类时，间接实现它。

Servlet 接口定义了 Servlet 与 Servlet 容器之间的契约。这个契约归结起来就是，Servlet 容器将 Servlet 类载入内存，并在 Servlet 实例上调用具体的方法。在一个应用程序中，每种 Servlet 类型只能有一个实例。

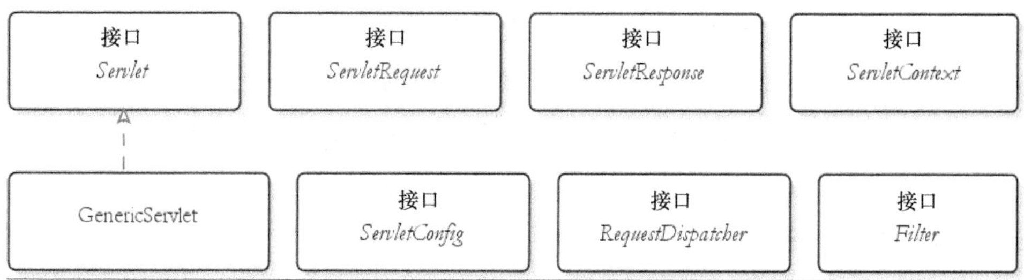

图 C.1 javax.servlet 中的主要类型

用户请求致使 Servlet 容器调用 Servlet 的 Service 方法，并传入一个 ServletRequest 实例和一个 ServletResponse 实例。ServletRequest 中封装了当前的 HTTP 请求，因此，Servlet 开发人员不必解析和操作原始的 HTTP 数据。ServletResponse 表示当前用户的 HTTP 响应，使得将响应发回给用户变得十分容易。

对于每一个应用程序，Servlet 容器还会创建一个 ServletContext 实例。这个对象中封装了上下文（应用程序）的环境详情。每个上下文只有一个 ServletContext。每个 Servlet 实例也都有一个封装 Servlet 配置的 ServletConfig。

下面来看 Servlet 接口。上面提到的其他接口，将在本章的其他小节中讲解。

C.2 Servlet

Servlet 接口中定义了以下 5 个方法：

```
void init(ServletConfig config) throws ServletException

void service(ServletRequest request, ServletResponse response)
        throws ServletException, java.io.IOException

void destroy()

java.lang.String getServletInfo()

ServletConfig getServletConfig()
```

注意，编写 Java 方法签名的惯例是，对于和包含该方法的类型不处于同一个包中的类型，要使用全类名。正因为如此，在 Service 方法 javax.servlet.ServletException 的签名中（与 Servlet 位于同一个包中），是没有包信息的，而 java.io.Exception 则是要写成完整名称的。

init、service 和 destroy 是生命周期方法。Servlet 容器根据以下规则调用这 3 个方法：

- init，当请求 Servlet 时，Servlet 容器会第一时间调用这个方法。这个方法在后续请求中不会再被调用。利用这个方法调用初始化代码。调用这个方法时，Servlet 容器会传入一个 ServletConfig。一般来说，你会将 ServletConfig 赋给一个类级变量，因此这个对象可以通过 Servlet 类的其他位置来使用。
- service，每当请求 Servlet 时，Servlet 容器就会调用这个方法。编写代码时，假设 Servlet 要在这里被请求。第一次请求 Servlet 时，Servlet 容器调用 init 方法和 Service 方法。后续的请求将只调用 service 方法。
- destroy，当要销毁 Servlet 时，Servlet 容器就会调用这个方法。当要卸载应用程序，或者当要关闭 Servlet 容器时，就会发生这种情况。一般会在这个方法中编写清除代码。

Servlet 中的另外两个方法是非生命周期方法，即 getServletInfo 和 getServletConfig。

- getServletInfo，这个方法会返回 Servlet 的描述。你可以返回有用或为 null 的任意字符串。
- getServletConfig，这个方法会返回由 Servlet 容器传给 init 方法的 ServletConfig。但是，为了让 getServletConfig 返回一个非 null 值，必须将传给 init 方法的 ServletConfig 赋给一个类级变量。ServletConfig 将在 C.6 小节中讲解。

注意线程安全性。Servlet 实例会被一个应用程序中的所有用户共享，因此不建议使用类级变量，除非它们是只读的，或者是 java.util.concurrent.atomic 包的成员。

C.3 将介绍如何编写 Servlet 实现。

C.3　编写基础的 Servlet 应用程序

其实，编写 Servlet 应用程序出奇简单。只需要创建一个目录结构，并把 Servlet 类放在某个目录下。本节将教你如何编写一个名为 servletapi 1 的 Servlet 应用程序。最初，它会包含一个 Servlet，即 MyServlet，其效果是向用户发出一条问候。

要运行 Servlets，还需要一个 Servlet 容器。Tomcat 是一个开源的 Servlet 容器，它是免费的，并且可以在任何能跑 Java 的平台上运行。如果你到现在都还没有安装 Tomcat，应该去看看附录 A，并安装一个。

C.3.1　编写和编译 Servlet 类

确定你的机器上有了 Servlet 容器后，下一步就要编写和编译一个 Servlet 类。本例中的 Servlet 类是 MyServlet，如清单 C.1 所示。按照惯例，Servlet 类的名称要以 Servlet 作为后缀。

清单 C.1　MyServlet 类

```
package servletapi1;
import java.io.IOException;
import java.io.PrintWriter;
import javax.servlet.Servlet;
import javax.servlet.ServletConfig;
import javax.servlet.ServletException;
import javax.servlet.ServletRequest;
import javax.servlet.ServletResponse;
import javax.servlet.annotation.WebServlet;

@WebServlet(name = "MyServlet", urlPatterns = { "/my" })
public class MyServlet implements Servlet {

    private transient ServletConfig servletConfig;

    @Override
    public void init(ServletConfig servletConfig)
            throws ServletException {
        this.servletConfig = servletConfig;
    }

    @Override
    public ServletConfig getServletConfig() {
        return servletConfig;
    }

    @Override
    public String getServletInfo() {
        return "My Servlet";
    }

    @Override
    public void service(ServletRequest request,
            ServletResponse response) throws ServletException,
            IOException {
        String servletName = servletConfig.getServletName();
        response.setContentType("text/html");
        PrintWriter writer = response.getWriter();
        writer.print("<!DOCTYPE html>"
```

```
                + "<html>"
                + "<body>Hello from " + servletName
                + "</body>"
                + "</html>");
    }

    @Override
    public void destroy() {
    }
}
```

看到清单 C.1 中的代码时，你可能首先注意到的是下面这个注解：

```
@WebServlet(name = "MyServlet", urlPatterns = { "/my" })
```

WebServlet 注解类型用来声明一个 Servlet。命名 Servlet 时，还可以暗示容器，是哪个 URL 调用这个 Servlet。name 属性是可选的，如有该属性，通常用 Servlet 类的名称。重要的是 urlPatterns 属性，它也是可选的，但是一般都是有的。在 MyServlet 中，urlPatterns 告诉容器，/my 样式表示应该调用 Servlet。

注意，URL 样式必须用一个正斜杆开头。

Servlet 的 init 方法只被调用一次，并将 private transient 变量 ServletConfig 设为传给该方法的 ServletConfig 对象。

```
private transient ServletConfig servletConfig;

@Override
public void init(ServletConfig servletConfig)
        throws ServletException {
    this.servletConfig = servletConfig;
}
```

如果想通过 Servlet 内部使用 ServletConfig，只需要将被传入的 ServletConfig 赋给一个类变量。

Service 方法发送字符串“Hello from MyServlet”给浏览器。对于每一个针对 Servlet 进来的 HTTP 请求，都会调用 Service 方法。

为了编译 Servlet，必须将 Servlet API 中的所有类型都放在你的类路径下。Tomcat 中带有 servlet-api.jar 文件，其中包含了 javax.servlet 的成员以及 javax.servlet.http 包。这个压缩文件放在 Tomcat 安装目录下的 lib 目录中。

C.3.2 应用程序目录结构

Servlet 应用程序必须在某一个目录结构下部署。图 C.2 展示了 Web 的应用程序目录。

web
META-INF
WEB-INF
classes
servletapi1
MyServlet.class

图 C.2 应用程序目录

这个目录结构最上面的那个 Web 目录，就是应用程序目录。在应用程序目录下，是 WEB-INF 目录。它有两个子目录：

- classes。Servlet 类及其他 Java 类必须放在这里面。类以下的目录反映了类包的结构。
- lib。Servlet 应用程序所需的 jar 文件要在这里部署。但 Servlet API 的 jar 文件不需要在这里部署，因为 Servlet 容器已经有它的备份。在这个应用程序中，lib 目录是空的。空的 lib 目录可以删除。

Servlet/JSP 应用程序一般都有 JSP 页面、HTML 文件、图片文件以及其他资料。这些应该放在应用程序目录下，并且经常放在子目录下。例如，所有的图片文件可以放在一个 image 目录下，所有的 JSP 页面可以放在 jsp 目录下，等等。

放在应用程序目录下的任何资源，用户只要输入资源 URL，都可以直接访问到。如果想让某一个资源可以被 Servlet 访问，但不可以被用户访问，那么就要把它放在 WEB-INF 目录下。

现在，准备将应用程序部署到 Tomcat。使用 Tomcat 时，一种部署方法是将应用程序目录复制到 Tomcat 安装目录下的 webapps 目录中。也可以通过在 Tomcat 的 conf 目录中编辑 server.xml 文件来实现部署，或者单独部署一个 XML 文件，这样就不需要编辑 server.xml 了。其他的 Servlet 容器可能会有不同的部署规则。关于如何将 Servlet/JSP 应用程序部署到 Tomcat 的详细信息，请查阅附录 A。

部署 Servlet/JSP 应用程序时，建议将它部署成一个 war 文件。war 文件其实就是以 war 作为扩展名的 jar 文件。利用带有 JDK 或者类似 WinZip 工具的 jar 软件，都可以创建 war 文件。然后，将 war 文件复制到 Tomcat 的 webapps 目录下。当开始启动 Tomcat 时，Tomcat 就会自动解压这个 war 文件。部署成 war 文件在所有 Servlet 容器中都适用。

C.3.3 调用 Servlet

要测试这个 Servlet，在浏览器中打开下面的 URL：

```
http://localhost:8080/servletapi1/my
```

其输出结果应该类似于图 C.3。

恭喜，你已经成功编写了第一个 Servlet 应用程序！

图 C.3 MyServlet 的响应

C.4　ServletRequest

对于每一个 HTTP 请求，Servlet 容器都会创建一个 ServletRequest 实例，并将它传给 Servlet 的 Service 方法。ServletRequest 封装了关于这个请求的信息。

ServletRequest 接口中有如下一些方法。

```
public int getContentLength()
```

返回请求主体的字节数。如果不知道字节长度，这个方法就会返回-1。

```
public java.lang.String getContentType()
```

返回请求主体的 MIME 类型，如果不知道类型，则返回 null。

```
public java.lang.String getParameter(java.lang.String name)
```

返回指定请求参数的值。

```
public java.lang.String getProtocol()
```

返回这个 HTTP 请求的协议名称和版本。

getParameter 是 ServletRequest 中最常用的方法。该方法通常用于返回 HTML 表单域的值。在 C.10 小节中，我们将会学到如何获取表单值。

getParameter 也可以用于获取查询字符串的值。例如，利用下面的 URI 调用 Servlet：

```
http://domain/context/servletName?id=123
```

用下面这个语句，可以通过 Servlet 内部获取 id 值：

```
String id = request.getParameter("id");
```

注意，如果该参数不存在，getParameter 将返回 null。

除了 getParameter 外，还可以使用 getParameterNames、getParameterMap 和 getParameterValues 获取表单域名、值以及查询字符串。这些方法的使用范例请参阅 C.9 小节。

C.5　ServletResponse

javax.servlet.ServletResponse 接口表示一个 Servlet 响应。在调用 Servlet 的 Service 方法前，

Servlet 容器首先创建一个 ServletResponse，并将它作为第 2 个参数传给 Service 方法。ServletResponse 隐藏了向浏览器发送响应的复杂过程。

在 ServletResponse 中定义的方法之一是 getWriter 方法，它返回了一个可以向客户端发送文本的 java.io.PrintWriter。默认情况下，PrintWriter 对象使用 ISO-8859-1 编码。

在向客户端发送响应时，大多数时候是将它作为 HTML 发送。因此，你必须非常熟悉 HTML。

注意：

还有一个方法可以用来向浏览器发送输出，它就是 getOutputStream。但这个方法是用于发送二进制数据的，因此，大多数情况使用的是 getWriter，而不是 getOutputStream。

在发送任何 HTML 标签前，应该先调用 setContentType 方法，设置响应的内容类型，并将“text/html”作为一个参数传入。这是在告诉浏览器，内容类型为 HTML。在没有内容类型的情况下，大多数浏览器会默认将响应渲染成 HTML。但是，如果没有设置响应内容类型，有些浏览器就会将 HTML 标签显示为普通文本。

C.6 ServletConfig

当 Servlet 容器初始化 Servlet 时，Servlet 容器会给 Servlet 的 init 方法传入一个 ServletConfig。ServletConfig 封装可以通过@WebServlet 或者部署描述符传给 Servlet 的配置信息。这样，传入的每一条信息就叫一个初始参数。一个初始参数有 key 和 value 两个元件。

为了从 Servlet 内部获取到初始参数的值，要在 Servlet 容器传给 Servlet 的 init 方法的 ServletConfig 中调用 getInitParameter 方法。getInitParameter 的方法签名如下：

```
java.lang.String getInitParameter(java.lang.String name)
```

此外，getInitParameterNames 方法则是返回所有初始参数名称的一个 Enumeration。

```
java.util.Enumeration<java.lang.String> getInitParameterNames()
```

例如，为了获取 contactName 参数值，要使用下面的方法签名：

```
String contactName = servletConfig.getInitParameter("contactName");
```

除 getInitParameter 和 getInitParameterNames 之外，ServletConfig 还提供了另一个很有用的方法：getServletContext。利用这个方法可以从 Servlet 内部获取 ServletContext。关于这个

对象的深入探讨，请查阅 C.7 小节。

下面举一个 ServletConfig 的范例，在 servletapi1 中添加一个名为 ServletConfigDemoServlet 的 Servlet。这个新的 Servlet 如清单 C.2 所示。

清单 C.2　ServletConfigDemoServlet 类

```
package servletapi1;
import java.io.IOException;
import java.io.PrintWriter;
import javax.servlet.Servlet;
import javax.servlet.ServletConfig;
import javax.servlet.ServletException;
import javax.servlet.ServletRequest;
import javax.servlet.ServletResponse;
import javax.servlet.annotation.WebInitParam;
import javax.servlet.annotation.WebServlet;

@WebServlet(name = "ServletConfigDemoServlet",
    urlPatterns = { "/servletConfigDemo" },
    initParams = {
        @WebInitParam(name="admin", value="Harry Taciak"),
        @WebInitParam(name="email", value="admin@example.com")
    }
)
public class ServletConfigDemoServlet implements Servlet {
    private transient ServletConfig servletConfig;

    @Override
    public ServletConfig getServletConfig() {
        return servletConfig;
    }

    @Override
    public void init(ServletConfig servletConfig)
            throws ServletException {
        this.servletConfig = servletConfig;
    }

    @Override
    public void service(ServletRequest request,
            ServletResponse response)
            throws ServletException, IOException {
        ServletConfig servletConfig = getServletConfig();
        String admin = servletConfig.getInitParameter("admin");
        String email = servletConfig.getInitParameter("email");
```

```
        response.setContentType("text/html");
        PrintWriter writer = response.getWriter();
        writer.print("<!DOCTYPE html>"
                + "<html>"
                + "<body>"
                + "Admin:" + admin
                + "<br/>Email:" + email
                + "</body></html>");
    }

    @Override
    public String getServletInfo() {
        return "ServletConfig demo";
    }

    @Override
    public void destroy() {
    }
}
```

如清单 C.2 所示，在@WebServlet 的 initParams 属性中，给 Servlet 传入了两个初始参数（admin 和 email）。

```
@WebServlet(name = "ServletConfigDemoServlet",
    urlPatterns = { "/servletConfigDemo" },
    initParams = {
        @WebInitParam(name="admin", value="Harry Taciak"),
        @WebInitParam(name="email", value="admin@example.com")
    }
)
```

用下面这个 URL，可以调用 ServletConfigDemoServlet：

```
http://localhost:8080/servletapi1/servletConfigDemo
```

其结果类似于图 C.4。

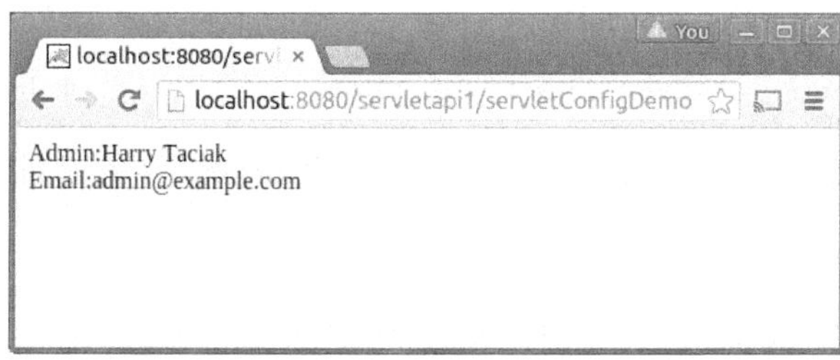

图 C.4 ServletConfigDemoServlet 效果展示

另一种方法是，在部署描述符中传入初始参数。在这里使用部署描述符，比使用@WebServlet更容易，因为部署描述符是一个文本文件，不需要重新编译Servlet类，就可以对它进行编辑。

部署描述符将在C.11节详细讲解。

C.7 ServletContext

ServletContext表示Servlet应用程序。每个Web应用程序只有一个上下文。在将一个应用程序同时部署到多个容器的分布式环境中时，每台Java虚拟机只有一个ServletContext对象。

通过在ServletConfig中调用getServletContext方法，可以获得ServletContext。

有了ServletContext，就可以共享可以从应用程序中的所有资料处访问到的信息，并且可以动态注册Web对象。前者将对象保存在ServletContext中的一个内部Map中。保存在ServletContext中的对象称作属性。

ServletContext中的下列方法负责处理属性：

```
java.lang.Object getAttribute(java.lang.String name)
java.util.Enumeration<java.lang.String> getAttributeNames()
void setAttribute(java.lang.String name, java.lang.Object object)
void removeAttribute(java.lang.String name)
```

C.8 GenericServlet

前面的例子展示了如何通过实现Servlet接口来编写Servlet。但你注意到没有？它们必须给Servlet中的所有方法都提供实现，即便其中有一些实现压根就没有包含任何代码。此外，还需要将ServletConfig对象保存到类级变量中。

值得庆幸的是，GenericServlet抽象类出现了。本着面向对象编程的尽可能使代码简单的原则，GenericServlet实现了Servlet和ServletConfig，并完成以下任务：

- 将init方法中的ServletConfig赋给一个类级变量，以便可以通过调用getServletConfig获取。
- 为Servlet接口中的所有方法提供默认的实现。
- 提供方法，包围ServletConfig中的方法。

GenericServlet 通过将 ServletConfig 赋给 init 方法中的类级变量 servletConfig，来保存 ServletConfig。下面就是 GenericServlet 中的 init 实现。

```
public void init(ServletConfig servletConfig)
        throws ServletException {
    this.servletConfig = servletConfig;
    this.init();
}
```

但是，如果在类中覆盖了这个方法，就会调用 Servlet 中的 init 方法，并且还必须调用 super.init(servletConfig)来保存 ServletConfig。为了避免上述麻烦，GenericServlet 提供了第 2 个 init 方法，它不带参数。这个方法是在 ServletConfig 被赋给 servletConfig 后，由第 1 个 init 方法调用。

```
public void init(ServletConfig servletConfig)
        throws ServletException {
    this.servletConfig = servletConfig;
    this.init();
}
```

这意味着，可以通过覆盖没有参数的 init 方法来编写初始化代码，ServletConfig 则仍然由 GenericServlet 实例保存。

清单 C.3 中的 GenericServletDemoServlet 类是对清单 C.2 中 ServletConfigDemoServlet 类的改写。注意，这个新的 Servlet 扩展了 GenericServlet，而不是实现 Servlet。

清单 C.3 GenericServletDemoServlet 类

```
package servletapi1;
import java.io.IOException;
import java.io.PrintWriter;
import javax.servlet.GenericServlet;
import javax.servlet.ServletConfig;
import javax.servlet.ServletException;
import javax.servlet.ServletRequest;
import javax.servlet.ServletResponse;
import javax.servlet.annotation.WebInitParam;
import javax.servlet.annotation.WebServlet;

@WebServlet(name = "GenericServletDemoServlet",
        urlPatterns = { "/generic" },
        initParams = {
            @WebInitParam(name="admin", value="Harry Taciak"),
            @WebInitParam(name="email", value="admin@example.com")
```

```
        }
)
public class GenericServletDemoServlet extends GenericServlet {

    private static final long serialVersionUID = 62500890L;

    @Override
    public void service(ServletRequest request,
            ServletResponse response)
            throws ServletException, IOException {
        ServletConfig servletConfig = getServletConfig();
        String admin = servletConfig.getInitParameter("admin");
        String email = servletConfig.getInitParameter("email");
        response.setContentType("text/html");
        PrintWriter writer = response.getWriter();
        writer.print("<!DOCTYPE html>"
                + "<html><head></head><body>"
                + "Admin:" + admin
                + "<br/>Email:" + email
                + "</body></html>");
    }
}
```

可见，通过扩展 GenericServlet，就不需要覆盖那些不打算改变的方法。因此，代码变得更加整洁。在清单 C.3 中，唯一被覆盖的方法是 Service 方法。而且，不必亲自保存 ServletConfig。

用下面这个 URL 调用 Servlet，其结果应该与 ServletConfigDemoServlet 相似。

```
http://localhost:8080/servletapi1/generic
```

即使 GenericServlet 是对 Servlet 一个很好的加强，但它也不常用，因为它毕竟不像 HttpServlet 那么高级。HttpServlet 才是主角，在现实的应用程序中广泛使用。关于它的详情，请查阅 C.9 节。

C.9　Http Servlets

即便不是全部，至少也是大多数应用程序都要与 HTTP 结合起来使用。这意味着可以利用 HTTP 提供的特性。javax.servlet.http 包是 Servlet API 中的第 2 个包，其中包含了用于编写 Servlet 应用程序的类和接口。javax.servlet.http 中的许多类型都覆盖了 javax.servlet 中的类型。

图 C.5 展示了 javax.servlet.http 中的主要类型。

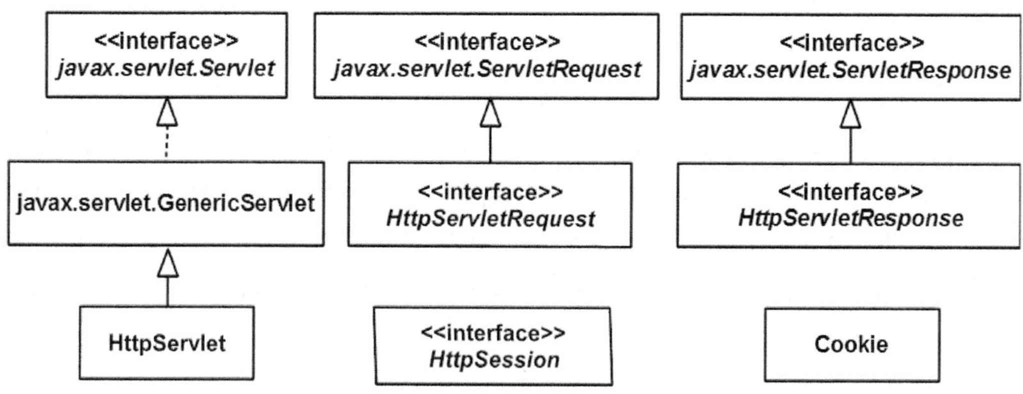

图 C.5 javax.servlet.http 中的主要类型

C.9.1 HttpServlet

HttpServlet 类覆盖了 javax.servlet.GenericServlet 类。使用 HttpServlet 时，还要借助分别代表 Servlet 请求和 Servlet 响应的 HttpServletRequest 和 HttpServletResponse 对象。HttpServletRequest 接口扩展 javax.servlet.ServletRequest，HttpServletResponse 扩展 javax.servlet.ServletResponse。

HttpServlet 覆盖 GenericServlet 中的 Service 方法，并通过下列签名再添加一个 Service 方法：

```
protected void service(HttpServletRequest request,
        HttpServletResponse response)
        throws ServletException, java.io.IOException
```

新 Service 方法和 javax.servlet.Servlet 中 Service 方法之间的区别在于，前者接受 HttpServletRequest 和 HttpServletResponse，而不是 ServletRequest 和 ServletResponse。

像往常一样，Servlet 容器调用 javax.servlet.Servlet 中原始的 Service 方法。HttpServlet 中的方法编写如下：

```
public void service(ServletRequest req, ServletResponse res)
        throws ServletException, IOException {
    HttpServletRequest request;
    HttpServletResponse response;
    try {
        request = (HttpServletRequest) req;
        response = (HttpServletResponse) res;
    } catch (ClassCastException e) {
        throw new ServletException("non-HTTP request or response");
    }
```

```
    service(request, response);
}
```

原始的Service方法将Servlet容器的request和response对象分别转换成HttpServletRequest和HttpServletResponse，并调用新的Service方法。这种转换总是会成功的，因为在调用Servlet的Service方法时，Servlet容器总会传入一个HttpServletRequest和一个HttpServletResponse，预备使用HTTP。即便正在实现javax.servlet.Servlet，或者扩展javax.servlet.GenericServlet，也可以将传给Service方法的servlet request和servlet response分别转换成HttpServletRequest和HttpServletResponse。

然后，HttpServlet中的Service方法会检验用来发送请求的HTTP方法（通过调用request.getMethod），并调用以下方法之一：doGet、doPost、doHead、doPut、doTrace、doOptions和doDelete。这7种方法中每一种方法都表示一个HTTP方法。doGet和doPost是最常用的。因此，不再需要覆盖Service方法了，只要覆盖doGet或者doPost，或者二者都覆盖即可。

总之，HttpServlet有两个特性在GenericServlet中尚未体现出来。

- 不覆盖Service方法，而是覆盖doGet或者doPost，或者覆盖doGet和doPost。在少数情况下，还会覆盖以下任意方法：doHead、doPut、doTrace、doOptions和doDelete。
- 与HttpServletRequest和HttpServletResponse共用，而不是与ServletRequest和ServletResponse共用。

HttpServletRequest表示HTTP环境中的Servlet请求。它扩展javax.servlet.ServletRequest接口，并添加了几个方法。新增的部分方法如下：

```
java.lang.String getContextPath()
```

返回表示请求上下文的请求RUI部分。

```
Cookie[] getCookies()
```

返回一个Cookie对象数组。

```
java.lang.String getHeader(java.lang.String name)
```

返回指定HTTP标题的值。

```
java.lang.String getMethod()
```

返回生成这个请求的HTTP方法名称。

```
java.lang.String getQueryString()
```

返回请求 URL 中的查询字符串。

```
HttpSession getSession()
```

返回与这个请求相关的会话对象。如果没有，将创建一个新的会话对象。

```
HttpSession getSession(boolean create)
```

返回与这个请求相关的会话对象。如果有，并且 create 参数为 True，将创建一个新的会话对象。

C.9.2 HttpServletResponse

HttpServletResponse 表示 HTTP 环境中的 Servlet 响应。下面是其中定义的部分方法：

```
void addCookie(Cookie cookie)
```

给这个响应对象添加一个 cookie。

```
void addHeader(java.lang.String name, java.lang.String value)
```

给这个响应对象添加一个 header。

```
void sendRedirect(java.lang.String location)
```

发送一条响应码，将浏览器跳转到指定的位置。

下面的章节将进一步学习这些方法。

C.10 处理 HTML 表单

一个 Web 应用程序中几乎总会包含一个或者多个 HTML 表单，供用户输入值。你可以轻松地将一个 HTML 表单从一个 Servlet 发送到浏览器。当用户提交表单时，在表单元素中输入的值就会被当作请求参数发送到服务器。

HTML 输入字段（文本字段、隐藏字段或者密码字段）或者文本区的值，会被当作字符串发送到服务器。空的输入字段或者文本域区会发送空的字符串。因此，有输入字段名称的 ServletRequest.getParameter 绝对不会返回 null。

HTML 的 select 元素也向 header 发送了一个字符串。如果 select 元素中没有任何选项被选中，那么就会发出所显示的这个选项值。

包含多个值的 select 元素（允许选择多个选项并且用<select multiple>表示的 select 元素）发出一个字符串数组，并且必须通过 SelectRequest.getParameterValues 进行处理。

复选框比较奇特。选中的复选框会发送字符串“on”到服务器。未选中的复选框则不向服务器发送任何内容，ServletRequest.getParameter(fieldName)返回 null。

单选框将被选中按钮的值发送到服务器。如果没有选择任何按钮，将没有任何内容被发送到服务器，并且 ServletRequest.getParameter(fieldName)返回 null。

如果一个表单中包含多个输入同名的元素，那么所有值都会被提交，并且必须利用 ServletRequest.getParameterValues 来获取它们。ServletRequest.getParameter 将只返回最后一个值。

清单 C.4 中的 FormServlet 类展示了如何处理 HTML 表单。它的 doGet 方法将一个 Order 表单发送到浏览器。它的 doPost 方法获取到所输入的值，并将它们输出。这个 Servlet 就是 servletapi2 应用程序的一部分。

清单 C.4　FormServlet 类

```
package servletapi2;
import java.io.IOException;
import java.io.PrintWriter;
import java.util.Enumeration;
import javax.servlet.ServletException;
import javax.servlet.annotation.WebServlet;
import javax.servlet.http.HttpServlet;
import javax.servlet.http.HttpServletRequest;
import javax.servlet.http.HttpServletResponse;

@WebServlet(name = "FormServlet", urlPatterns = { "/form" })
public class FormServlet extends HttpServlet {
    private static final long serialVersionUID = 54L;
    private static final String TITLE = "Order Form";

    @Override
    public void doGet(HttpServletRequest request,
            HttpServletResponse response)
            throws ServletException, IOException {
        response.setContentType("text/html");
        PrintWriter writer = response.getWriter();
        writer.println("<!DOCTYPE html>");
        writer.println("<html>");
        writer.println("<head>");
        writer.println("<title>" + TITLE + "</title></head>");
        writer.println("<body><h1>" + TITLE + "</h1>");
```

```
writer.println("<form method='post'>");
writer.println("<table>");
writer.println("<tr>");
writer.println("<td>Name:</td>");
writer.println("<td><input name='name'/></td>");
writer.println("</tr>");
writer.println("<tr>");
writer.println("<td>Address:</td>");
writer.println("<td><textarea name='address' "
        + "cols='40' rows='5'></textarea></td>");
writer.println("</tr>");
writer.println("<tr>");
writer.println("<td>Country:</td>");
writer.println("<td><select name='country'>");
writer.println("<option>United States</option>");
writer.println("<option>Canada</option>");
writer.println("</select></td>");
writer.println("</tr>");
writer.println("<tr>");
writer.println("<td>Delivery Method:</td>");
writer.println("<td><input type='radio' " +
        "name='deliveryMethod'"
        + " value='First Class'/>First Class");
writer.println("<input type='radio' " +
        "name='deliveryMethod' "
        + "value='Second Class'/>Second Class</td>");
writer.println("</tr>");
writer.println("<tr>");
writer.println("<td>Shipping Instructions:</td>");
writer.println("<td><textarea name='instruction' "
        + "cols='40' rows='5'></textarea></td>");
writer.println("</tr>");
writer.println("<tr>");
writer.println("<td> </td>");
writer.println("<td><textarea name='instruction' "
        + "cols='40' rows='5'></textarea></td>");
writer.println("</tr>");
writer.println("<tr>");
writer.println("<td>Please send me the latest " +
        "product catalog:</td>");
writer.println("<td><input type='checkbox' " +
        "name='catalogRequest'/></td>");
writer.println("</tr>");
writer.println("<tr>");
writer.println("<td> </td>");
writer.println("<td><input type='reset'/>" +
```

```
                "<input type='submit'/></td>");
        writer.println("</tr>");
        writer.println("</table>");
        writer.println("</form>");
        writer.println("</body>");
        writer.println("</html>");
    }

    @Override
    public void doPost(HttpServletRequest request,
            HttpServletResponse response)
            throws ServletException, IOException {
        response.setContentType("text/html");
        PrintWriter writer = response.getWriter();
        writer.println("<html>");
        writer.println("<head>");
        writer.println("<title>" + TITLE + "</title></head>");
        writer.println("</head>");
        writer.println("<body><h1>" + TITLE + "</h1>");
        writer.println("<table>");
        writer.println("<tr>");
        writer.println("<td>Name:</td>");
        writer.println("<td>" + request.getParameter("name")
                + "</td>");
        writer.println("</tr>");
        writer.println("<tr>");
        writer.println("<td>Address:</td>");
        writer.println("<td>" + request.getParameter("address")
                + "</td>");
        writer.println("</tr>");
        writer.println("<tr>");
        writer.println("<td>Country:</td>");
        writer.println("<td>" + request.getParameter("country")
                + "</td>");
        writer.println("</tr>");
        writer.println("<tr>");
        writer.println("<td>Shipping Instructions:</td>");
        writer.println("<td>");
        String[] instructions = request
                .getParameterValues("instruction");
        if (instructions != null) {
            for (String instruction : instructions) {
                writer.println(instruction + "<br/>");
            }
        }
        writer.println("</td>");
```

```
        writer.println("</tr>");
        writer.println("<tr>");
        writer.println("<td>Delivery Method:</td>");
        writer.println("<td>"
                + request.getParameter("deliveryMethod")
                + "</td>");
        writer.println("</tr>");
        writer.println("<tr>");
        writer.println("<td>Catalog Request:</td>");
        writer.println("<td>");
        if (request.getParameter("catalogRequest") == null) {
            writer.println("No");
        } else {
            writer.println("Yes");
        }
        writer.println("</td>");
        writer.println("</tr>");
        writer.println("</table>");
        writer.println("<div style='border:1px solid #ddd;" +
                    "margin-top:40px;font-size:90%'>");

        writer.println("Debug Info<br/>");
        Enumeration<String> parameterNames = request
                .getParameterNames();
        while (parameterNames.hasMoreElements()) {
            String paramName = parameterNames.nextElement();
            writer.println(paramName + ": ");
            String[] paramValues = request
                    .getParameterValues(paramName);
            for (String paramValue : paramValues) {
                writer.println(paramValue + "<br/>");
            }
        }
        writer.println("</div>");
        writer.println("</body>");
        writer.println("</html>");
    }
}
```

用下面的 URL，可以调用 FormServlet：

```
http://localhost:8080/servletapi2/form
```

被调用的 doGet 方法会由这个 HTML 表单发送给浏览器。

```
<form method='post'>
<input name='name'/>
```

```
    <textarea name='address' cols='40' rows='5'></textarea>
    <select name='country'>");
        <option>United States</option>
        <option>Canada</option>
    </select>
    <input type='radio' name='deliveryMethod' value='First Class'/>
    <input type='radio' name='deliveryMethod' value='Second Class'/>
    <textarea name='instruction' cols='40' rows='5'></textarea>
    <textarea name='instruction' cols='40' rows='5'></textarea>
    <input type='checkbox' name='catalogRequest'/>
    <input type='reset'/>
    <input type='submit'/>
    </form>
```

表单的方法设为 post，确保当用户提交表单时，使用 HTTP POST 方法。它的 action 属性缺省，表示该表单会被提交给请求它时使用的相同的 URL。

图 C.6 展示了一个空的 Order 表单。

现在，填写表单，并单击 Submit 按钮。在表单中输入的值，将利用 HTTP POST 方法发送给服务器，这样就会调用 Servlet 的 doPost 方法。因此，你将会看到如图 C.7 所示的那些值。

图 C.6　一个空的 Order 表单

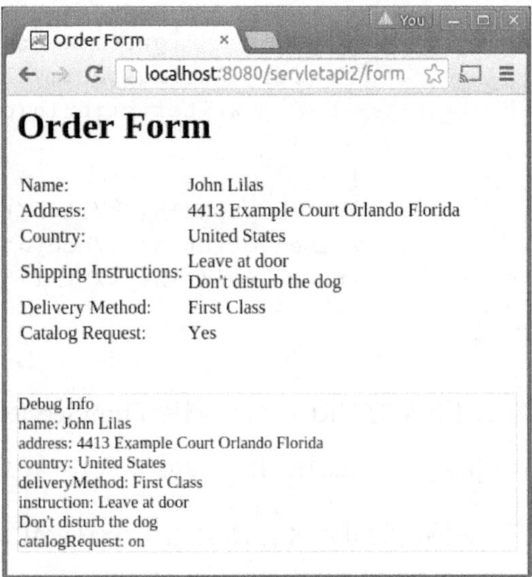

图 C.7　在 Order 表单中输入的值

C.11 使用部署描述符

正如在前面的例子中所见到的，编写和部署 Servlet 都是很容易的事情。部署的一个方面是用一个路径配置 Servlet 的映射。在这些范例中，是利用 WebServlet 标注类型，用一个路径映射了一个 Servlet。

利用部署描述符是配置 Servlet 应用程序的另一种方法，部署描述符总是命名为 web.xml，并且放在 WEB-INF 目录下。本章介绍了如何创建一个名为 servletapi3 的 Servlet 应用程序，并为它编写了一个 web.xml。

servletapi3 有 SimpleServlet 和 WelcomeServlet 两个 Servlet，还有一个要映射 Servlets 的部署描述符。清单 C.5 和清单 C.6 分别展示了 SimpleServlet 和 WelcomeServlet。注意，Servlet 类是没有用@WebServlet 注解的。部署描述符如清单 C.7 所示。

清单 C.5　未注解的 SimpleServlet 类

```
package servletapi3;
import java.io.IOException;
import java.io.PrintWriter;
import javax.servlet.ServletException;
import javax.servlet.http.HttpServlet;
import javax.servlet.http.HttpServletRequest;
import javax.servlet.http.HttpServletResponse;

public class SimpleServlet extends HttpServlet {
    private static final long serialVersionUID = 8946L;

    @Override
    public void doGet(HttpServletRequest request,
            HttpServletResponse response)
            throws ServletException, IOException {
        response.setContentType("text/html");
        PrintWriter writer = response.getWriter();
        writer.print("<!DOCTYPE html><html><head></head>" +
                "<body>Simple Servlet</body></html");
    }
}
```

清单 C.6　未注解的 WelcomeServlet 类

```
package servletapi3;
```

```
import java.io.IOException;
import java.io.PrintWriter;
import javax.servlet.ServletException;
import javax.servlet.http.HttpServlet;
import javax.servlet.http.HttpServletRequest;
import javax.servlet.http.HttpServletResponse;

public class WelcomeServlet extends HttpServlet {
    private static final long serialVersionUID = 27126L;

    @Override
    public void doGet(HttpServletRequest request,
            HttpServletResponse response)
            throws ServletException, IOException {
        response.setContentType("text/html");
        PrintWriter writer = response.getWriter();
        writer.print("<!DOCTYPE html><html><head></head>"
                + "<body>Welcome</body></html>");
    }
}
```

清单 C.7 部署描述符

```
<?xml version="1.0" encoding="UTF-8"?>
<web-app version="3.1"
    xmlns="http://xmlns.jcp.org/xml/ns/javaee"
    xmlns:xsi="http://www.w3.org/2001/XMLSchema-instance"
    xsi:schemaLocation="http://xmlns.jcp.org/xml/ns/javaee
➥ http://xmlns.jcp.org/xml/ns/javaee/web-app_3_1.xsd">

    <servlet>
        <servlet-name>SimpleServlet</servlet-name>
        <servlet-class>servletapi3.SimpleServlet</servlet-class>
        <load-on-startup>10</load-on-startup>
    </servlet>

    <servlet-mapping>
        <servlet-name>SimpleServlet</servlet-name>
        <url-pattern>/simple</url-pattern>
    </servlet-mapping>


    <servlet>
        <servlet-name>WelcomeServlet</servlet-name>
        <servlet-class>servletapi3.WelcomeServlet</servlet-class>
        <load-on-startup>20</load-on-startup>
```

```
    </servlet>

    <servlet-mapping>
        <servlet-name>WelcomeServlet</servlet-name>
        <url-pattern>/welcome</url-pattern>
    </servlet-mapping>
</web-app>
```

使用部署描述符有诸多好处。其一，可以将没有对等元素的元素放在@WebServlet 中，如 load-on-startup 元素。这个元素在应用程序启动时加载 Servlet，而不是在第一次调用 Servlet 时加载。使用 load-on-startup 意味着第一次调用 Servlet 所花的时间并不比后续的调用长。如果 Servlet 的 init 方法需要花费一些时间才能完成的话，这项功能就特别有用。

使用部署描述符的另一个好处是，如果需要修改配置值，如 Servlet 路径，则不需要重新编译 Servlet 类。

此外，可以将初始参数传给一个 Servlet，并且不需要重新编译 Servlet 类，就可以对它们进行编辑。

部署描述符还允许覆盖在 Servlet 标注中定义的值。Servlet 上的 WebServlet 注解如果同时也在部署描述符中进行声明，那么它将不起作用。但是，在有部署描述符的应用程序中，却不在部署描述符中注解的 Servlet 时，则仍然有效。这意味着，可以注解 Servlet，并在同一个应用程序的部署描述符中声明这些 Servlet。

图 C.8 展示了有部署描述符的 servletapi3 的目录结构。这个目录结构与 servletapi1 的目录结构没有太大区别。唯一的区别在于，servletapi3 在 WEB-INF 目录中有一个 web.xml 文件（部署描述符）。

图 C.8 有部署描述符的 servletapi3 的目录结构

现在，在部署描述符中声明 SimpleServlet 和 WelcomeServlet，可以利用这些 URL 来访问它们：

```
http://localhost:8080/servletapi3/simple
http://localhost:8080/servletapi3/welcome
```

C.12 小结

Servlet 技术是 Java EE 的一部分。所有 Servlet 都在 Servlet 容器中运行，容器和 Servlet 之间的契约采用 javax.servlet.Servlet 接口的形式。javax.servlet 包还提供实现 Servlet 的 GenericServlet 抽象类。这是一个方便的类，你可以扩展以创建一个 Servlet。然而，大多数现代 Servlet 将在 HTTP 环境中工作。因此，对 javax.servlet.http.HttpServlet 类进行子类化更有意义。HttpServlet 类本身是 GenericServlet 的子类。

附录D JavaServer Pages

Servlet 有两个缺点是无法克服的：首先，写在 Servlet 中的所有 HTML 标签必须包含 Java 字符串，这使得处理 HTTP 响应报文的工作十分繁琐；第二，所有的文本和 HTML 标记是硬编码，导致即使是表现层的微小变化，如改变背景颜色，也需要重新编译。

JavaServer Pages（JSP）解决了上述两个问题。同时，JSP 不会取代 Servlet，相反，它们具有互补性。现代的 Java Web 应用会同时使用 Servlet 和 JSP 页面。撰写本书时，JSP 的最新版本是 2.3。

本附录概述了 JSP 技术，并讨论了在 JSP 页面中，隐式对象以及 3 个语法元素（指令、脚本元素和动作），还讨论了错误处理。

注意

一种好设计方法是：不要在 JSP 页面中编写 Java 代码。除非应用程序只包含一个或两个简单的 JSP 页面，并且永远不会增长；否则，你应该采用模型 2，因为它规定 JSP 页面仅用于显示 Java 对象中的值。

D.1 JSP 概述

JSP 页面本质上是一个 Servlet。然而，用 JSP 页面开发比使用 Servlet 更容易，主要有两个原因。首先，不必编译 JSP 页面；其次，JSP 页面是一个以 jsp 为扩展名的文本文件，可以使用任何文本编辑器来编写它们。

JSP 页面在 JSP 容器中运行，一个 Servlet 容器通常也是 JSP 容器。例如，Tomcat 就是一个 Servlet/JSP 容器。

当一个 JSP 页面第一次被请求时，Servlet/JSP 容器主要做以下两件事情：

（1）把 JSP 页面转换到 JSP 页面实现类，该实现类是一个实现 javax.servlet.jsp.JspPage 接口或子接口 javax.servlet.jsp.HttpJspPage 的 Java 类。JspPage 是 javax.servlet.Servlet 的子接口，这使得每一个 JSP 页面都是一个 Servlet。该实现类的类名由 Servlet/JSP 容器生成。如果出现转换错误，则相关错误信息将被发送到客户端。

（2）如果转换成功，Servlet/JSP 容器随后编译该 Servlet 类，并装载和实例化该类，像其他正常的 Servlet 一样执行生命周期操作。

对于同一个 JSP 页面的后续请求，Servlet/JSP 容器会先检查 JSP 页面是否被修改过。如果是，则该 JSP 页面会被重新转换、编译并执行。如果不是，则执行已经在内存中的 JSP Servlet。这样一来，一个 JSP 页面的第一次调用的实际花费总比后来的花费多，因为它涉及转换和编译。为了解决这个问题，可以执行下列动作之一：

- 配置应用程序，使所有的 JSP 页面在应用程序启动时被调用（实际上也可视为转换和编译），而不是在第一次请求时调用。
- 预编译 JSP 页面，并将其部署为 Servlet。

JSP 自带的 API 包含 4 个包：

- javax.servlet.jsp。包含 Servlet/JSP 容器用于将 JSP 页面转换为 Servlet 的核心类和接口。其中的两个重要成员是 JspPage 和 HttpJspPage 接口。所有的 JSP 页面实现类必须实现 JspPage 或 HttpJspPage 接口。在 HTTP 环境下，实现 HttpJspPage 接口是显而易见的选择。
- javax.servlet.jsp.tagext。包括用于开发自定义标签的类型。
- javax.el。提供了统一表达式语言的 API。
- javax.servlet.jsp.el。提供了一组必须由 Servlet/JSP 容器支持以便在 JSP 页面中使用表达式语言的类。

除了 javax.servlet.jsp.tagext，我们很少直接使用 JSP API。事实上，编写 JSP 页面时，我们更关心 Servlet API，而非 JSP API。当然，我们还需要掌握 JSP 语法，本附录后续会进一步说明。JSP API 在开发 JSP 容器或 JSP 编译器时被广泛使用。

可以在以下网址查看 JSP API：

```
https://docs.oracle.com/javaee/7/api/javax/servlet/jsp/ package-summary.html
```

JSP 页面可以包含模板数据和语法元素。这里，语法元素是一些具有特殊意义的 JSP 转换符。例如，“<%” 是一个元素，因为它表示在 JSP 页面中的 Java 代码块的开始。“%>” 也是一个元素，因为它是 Java 代码块的结束符。除语法元素之外的一切都是模板数据。模板数据会原样发送给浏览器。例如，JSP 页面中的 HTML 标记和文字都是模板数据。

清单 D.1 给出一个名为 welcome.jsp 的 JSP 页面。它是发送一个客户问候的简单页面。注意，同 Servlet 相比，JSP 页面是如何更简单地完成同样的事情的。

清单 D.1 welcome.jsp

```
<!DOCTYPE html>
<html>
<head><title>Welcome</title></head>
<body>
Welcome
</body>
</html>
```

在 Tomcat 中，welcome.jsp 页面在第一次请求时被转换成名为 welcome_jsp 的 Servlet。你可以在 Tomcat 工作目录下的子目录中找到生成的 Servlet，该 Servlet 继承自 org.apache.jasper.runtime.HttpJspBase，这是一个抽象类，它继承自 javax.servlet.http.HttpServlet 并实现了 javax.servlet.jsp.HttpJspPage。

下面是为 welcome.jsp 生成的 Servlet。如果觉得不好理解，可以先跳过它。当然，能够理解它更好。

```
package org.apache.jsp;
import javax.servlet.*;
import javax.servlet.http.*;
import javax.servlet.jsp.*;

public final class welcome_jsp extends
        org.apache.jasper.runtime.HttpJspBase
        implements org.apache.jasper.runtime.JspSourceDependent {

    private static final javax.servlet.jsp.JspFactory _jspxFactory =
        javax.servlet.jsp.JspFactory.getDefaultFactory();

    private static java.util.Map<java.lang.String,java.lang.Long>
        _jspx_dependants;

    private javax.el.ExpressionFactory _el_expressionfactory;
    private org.apache.tomcat.InstanceManager _jsp_instancemanager;
```

```
public java.util.Map<java.lang.String,java.lang.Long>
    getDependants() {
        return _jspx_dependants;
}

public void _jspInit() {
    _el_expressionfactory =
            _jspxFactory.getJspApplicationContext(
            getServletConfig().getServletContext())
           .getExpressionFactory();
    _jsp_instancemanager =
            org.apache.jasper.runtime.InstanceManagerFactory
            .getInstanceManager(getServletConfig());
}

public void _jspDestroy() {
}

public void _jspService(final
    javax.servlet.http.HttpServletRequest request, final
    javax.servlet.http.HttpServletResponse response)
    throws java.io.IOException, javax.servlet.ServletException {

    final javax.servlet.jsp.PageContext pageContext;
    javax.servlet.http.HttpSession session = null;
    final javax.servlet.ServletContext application;
    final javax.servlet.ServletConfig config;
    javax.servlet.jsp.JspWriter out = null;
    final java.lang.Object page = this;
    javax.servlet.jsp.JspWriter _jspx_out = null;
    javax.servlet.jsp.PageContext _jspx_page_context = null;

    try {
        response.setContentType("text/html");
        pageContext = _jspxFactory.getPageContext(this, request,
            response, null, true, 8192, true);
        _jspx_page_context = pageContext;
        application = pageContext.getServletContext();
        config = pageContext.getServletConfig();
        session = pageContext.getSession();
        out = pageContext.getOut();
        _jspx_out = out;

        out.write("<html>\n");
        out.write("<head><title>Welcome</title></head>\n");
        out.write("<body>\n");
```

```
                out.write("Welcome\n");
                out.write("</body>\n");
                out.write("</html>");
            } catch (java.lang.Throwable t) {
                if (!(t instanceof
                        javax.servlet.jsp.SkipPageException)){
                    out = _jspx_out;
                    if (out != null && out.getBufferSize() != 0)
                        try {
                            out.clearBuffer();
                        } catch (java.io.IOException e) {
                        }
                    if (_jspx_page_context != null)
                       _jspx_page_context.handlePageException(t);
                }
            } finally {
                _jspxFactory.releasePageContext(_jspx_page_context);
            }
        }
    }
```

正如我们在上面的代码中看到的，JSP 页面的主体是_jspService 方法。这个方法被定义在 HttpJspPage 中，并被 HttpJspBase 的 service 方法调用。下面的代码来自 HttpJspBase 类。

```
public final void service(HttpServletRequest request,
        HttpServletResponse response) throws ServletException,
        IOException {
    _jspService(request, response);
}
```

要覆盖 init 和 destroy 方法，可以参见 D.5。

一个 JSP 页面不同于一个 Servlet 的另一个方面是，前者不需要添加注解或在部署描述符配置映射 URL。应用程序目录中的每一个 JSP 页面可以直接在浏览器中输入路径页面来访问。图 D.1 给出了 D1 应用程序的目录结构

web
META-INF
WEB-INF
welcome.jsp

图 D.1 D1 应用程序的目录结构

D1 应用程序的结构非常简单，由一个空的 WEB-INF 目录和 welcome.jsp 页面构成。

可以通过如下 URL 访问 welcome.jsp 页面。

```
http://localhost:8080/jspdemo/welcome.jsp
```

注意

添加新的 JSP 界面后，无需重启 Tomcat。

清单 D.2 展示了如何在 JSP 页面中使用 Java 代码来生成动态页面。清单 D.2 中的 todaysDate.jsp 页面显示了今天的日期。

清单 D.2　todaysDate.jsp 页面

```
<%@page import="java.util.Date"%>
<%@page import="java.text.DateFormat"%>
<!DOCTYPE html>
<html>
<head><title>Today's date</title></head>
<body>
<%
    DateFormat dateFormat =
            DateFormat.getDateInstance(DateFormat.LONG);
    String s = dateFormat.format(new Date());
    out.println("Today is " + s);
%>
</body>
</html>
```

todaysDate.jsp 页面将几个 HTML 标签和字符串"今天是"以及今天的日期发送到浏览器。

请注意两件事情。首先，Java 代码可以出现在 JSP 页面中的任何位置，并通过"<%"和"%>"包括起来。其次，可以使用 page 指令的 import 属性导入在 JSP 页面中使用的 Java 类型，如果没有导入的类型，必须在代码中写出 Java 类的全路径名称。

<%...%>块称为 scriplet，并在 D.5 部分进一步讨论。page 将在 C.4 部分详细讨论。

现在可以通过如下 URL 访问 todaysDate.jsp 页面。

```
http://localhost:8080/jspdemo/todaysDate.jsp
```

D.2　注释

为 JSP 页面添加注释是一个良好的习惯。JSP 支持两种不同的注释格式。

（1）JSP 注释。该注释记录页面中做了什么。

（2）HTML/XHTML 注释。这些注释将会发送到浏览器。

JSP 注释以"<%--"开始，以"--%>"结束。下面是一个例子：

```
<%-- retrieve products to display --%>
```

JSP 注释不会被发送到浏览器端，也不会被嵌套。

HTML/XHTML 注释语法如下：

```
<!-- [comments here] -->
```

一个 HTML/XHTML 注释不会被容器处理，会原样发送给浏览器。HTML/XHTML 注释的一个用途是用来确定 JSP 页面本身。

```
<!-- this is /jsp/store/displayProducts.jspf -->
```

当一个应用中存在多个 JSP 片段时，这会特别有用。开发人员可以很容易地通过在浏览器中查看 HTML 源代码来找出是哪一个 JSP 页面或片段产生了相应的 HTML 片段。

D.3 隐式对象

Servlet 容器会传递几个对象给它运行的 Servlet。例如，可以通过 Servlet 的 service 方法获取 HttpServletRequest 和 HttpServletResponse 对象，并且可以通过 init 方法访问到 ServletConfig 对象。此外，可以通过调用 HttpServletRequest 对象的 getSession 方法访问到 HttpSession 对象。

在 JSP 中，可以通过使用隐式对象来访问上述对象。表 D.1 展示了 JSP 隐式对象。

表 D.1 JSP 隐式对象

对象	类型
request	javax.servlet.http.HttpServletRequest
response	javax.servlet.http.HttpServletResponse
out	javax.servlet.jsp.JspWriter
session	javax.servlet.http.HttpSession
application	javax.servlet.ServletContext
config	javax.servlet.ServletConfig
pageContext	javax.servlet.jsp.PageContext
page	javax.servlet.jsp.HttpJspPage
exception	java.lang.Throwable

以 request 为例，该隐式对象代表 Servlet/JSP 容器传递给 Servlet 服务方法的 HttpServletRequest 对象。可以将 request 理解为指向 HttpServletRequest 对象的一个引用变量。下面的代码示例，

从 HttpServletRequest 对象中返回 username 参数值。

```
<%
    String userName = request.getParameter("userName");
%>
```

pageContext 用于 javax.servlet.jsp.PageContext。它提供了有用的上下文信息，并通过其自说明的方法来访问各种 Servlet 相关对象，如 getRequest、getResponse、getServletContext、getServletConfig 和 getSession。当然，这些方法在脚本中不是非常有用，因为可以更直接地通过隐式对象来访问 request、response、session 和 application。

此外，PageContext 提供了另一组有趣的方法：用于获取和设置属性的方法，即 getAttribute 方法和 setAttribute 方法。属性值可被存储在 4 个范围之一：页面、请求、会话和应用程序。页面范围是最小范围，这里存储的属性只在同一个 JSP 页面可用。请求范围是指当前的 ServletRequest。会话范围指当前的 HttpSession 中。应用程序范围是指应用的 ServletContext 中。

PageContext 的 setAttribute 方法签名如下：

```
public abstract void setAttribute(java.lang.String name,
        java.lang.Object value, int scope)
```

其中，scope 的取值范围为 PageContext 对象的最终静态 int 值：PAGE_SCOPE、REQUEST_SCOPE、SESSION_SCOPE 和 APPLICATION_SCOPE。

若要保存一个属性到页面范围，可以直接使用 setAttribute 重载方法。

```
public abstract void setAttribute(java.lang.String name,
        java.lang.Object value)
```

如下脚本将一个属性保存到 ServletRequest 中。

```
<%
    //product is a Java object
    pageContext.setAttribute("product", product,
            PageContext.REQUEST_SCOPE);
%>
```

同样效果的 Java 代码如下:

```
<%
    request.setAttribute("product", product);
%>
```

隐式对象 out 引用了一个 javax.servlet.jsp.JspWriter 对象，这类似于你在调用 HttpServletResponse 的 getWriter 方法时得到 java.io.PrintWriter。可以通过调用它的 print 方法将消息发送到浏览

器。例如：

```
out.println("Welcome");
```

清单 D.3 中的 implicitObjects.jsp 页面展示了部分隐式对象的使用。

清单 D.3　implicitObjects.jsp 页面

```
<%@page import="java.util.Enumeration"%>
<!DOCTYPE html>
<html>
<head><title>JSP Implicit Objects</title></head>
<body>
<b>Http headers:</b><br/>
<%
    for (Enumeration<String> e = request.getHeaderNames();
            e.hasMoreElements(); ){
        String header = e.nextElement();
        out.println(header + ": " + request.getHeader(header) +
                "<br/>");
    }
%>
<hr/>
<%
    out.println("Buffer size: " + response.getBufferSize() +
        "<br/>");
    out.println("Session id: " + session.getId() + "<br/>");
    out.println("Servlet name: " + config.getServletName() +
        "<br/>");
    out.println("Server info: " + application.getServerInfo());
%>
</body>
</html>
```

可以通过访问如下 URL 来调用 implicitObjects.jsp 页面：

```
http://localhost:8080/jspdemo/implicitObjects.jsp
```

该页面产生了如下内容：

```
Http headers:
host: localhost:8080
connection: keep-alive
accept:text/html,application/xhtml+xml,application/xml;q=0.9,image/webp,
➥ */*;q=0.8
user-agent: Mozilla/5.0 (X11; Linux x86_64) AppleWebKit/537.36 (KHTML,
➥ like Gecko) Chrome/43.0.2357.130 Safari/537.36
accept-encoding: gzip, deflate, sdch
```

```
accept-language: en-us ,en;q=0.8,id;q=0.6,ms;q=0.4,fr;q=0.2,de;q=0.2
cookie: JSESSIONID=4E3D1A4994B7F7ED5D7B96C2E3CF3BDE;
Buffer size: 8192
Session id: A4EC1E2FFCE1377FD6DF6545BDDD7909
Servlet name: jsp
Server info: Apache Tomcat/8.0.20
```

在浏览器中具体看到的内容，取决于所使用的浏览器及其环境。

注意，在默认情况下，JSP 编译器会将 JSP 页面的内容类型设为 text/html。如果要使用不同的类型，则需要调用 response.setContentType()或者使用页面指令（详情请参考 D.4 小节）来设置内容类型。例如，将内容类型设置为 text/json 如下所示：

```
response.setContentType("text/json");
```

还要注意的是，页面隐式对象是表示当前的 JSP 页面，JSP 页面的设计者一般不使用它。

D.4　指令

指令是 JSP 语法元素的第一种类型。它们指示 JSP 转换器如何把 JSP 页面转换为 Servlet。JSP 2.3 定义了多个指令，但只有 page 和 include 是最重要的，本节会详细讨论。

D.4.1　page 指令

可以使用 page 指令来控制 JSP 转换器转换当前 JSP 页面的某些方面。例如，可以告诉 JSP 用于转换隐式对象 out 的缓冲器的大小、内容类型，以及需要导入的 Java 类型，等等。

page 指令的语法如下：

```
<%@ page attribute1="value1" attribute2="value2" ... %>
```

@和 page 间的空格不是必需的，attribute1、attribute2 等是 page 指令的属性。如下是 page 指令属性的列表。

- import：定义一个或多个本页面中将被导入和使用的 java 类型。例如：import="java.util.List" 将导入 List 接口。可以使用通配符“*”来引入整个包，类似 import="java.util.*"。可以通过在两个类型间加入“,”分隔符来导入多个类型，如 import="java.util.ArrayList, java.util.Calendar, java.io.PrintWriter"。此外，JSP 默认导入如下包:java.lang、javax.servlet、javax.servlet.http、javax.servlet.jsp。

- session：值为 True，本页面加入会话管理；值为 False 则相反。默认值为 True，访问该页面时，若当前不存在 javax.servlet.http.HttpSession 实例，则会创建一个。
- buffer：以 kB 为单位，定义隐式对象 out 的缓冲大小。必须以 kB 后缀结尾。默认大小为 8kB 或更大（取决于 JSP 容器）。该值可以为 none，这意味着没有缓冲，所有数据将直接写入 PrintWriter。
- autoFlush：默认值为 True。若值为 True，则当输出缓冲满时会自写入输出流。而值为 False，则仅当调用隐式对象的 flush 方法时，才会写入输出流。因此，若缓冲溢出，则会抛出异常。
- isThreadSafe：定义该页面的线程安全级别。不推荐使用 JSP 参数，因为使用该参数后，会生成一些 Servlet 容器已过期的代码。
- info：指定生成的 Servlet 类的 getServletInfo 方法的返回值。
- errorPage：定义当出错时用来处理错误的页面。
- isErrorPage：标识本页是一个错误处理页面。
- contentType：定义本页面隐式对象 response 的内容类型，默认是 text/html。
- pageEncoding：定义本页面的字符编码，默认是 ISO-8859-1。
- isELIgnored：配置是否忽略 EL 表达式。EL 是 Expression Language 的缩写。
- language：定义本页面的脚本语言类型，默认是 Java，这在 JSP 2.2 中是唯一的合法值。
- extends：定义 JSP 实现类要继承的父类。这个属性较少使用，仅在非常特殊理由下使用。
- deferredSyntaxAllowedAsLiteral：定义是否解析字符串中出现“#{”符号，默认是 False。“{# ”是一个表达式语言的起始符号，因而很重要。
- trimDirectiveWhitespaces：定义是否不输出多余的空格/空行，默认是 False。

大部分 page 指令可以出现在页面的任何位置，但当 page 指令包含 contentType 或 pageEncoding 属性时，其必须出现在 Java 代码发送任何内容之前。这是因为内容类型和字符编码必须在发送任何内容前设定。

page 指令也可以出现多次，但出现多次的指令属性必须具有相同的值。不过，import 属性例外，多个包含 import 属性的 page 指令的结果是累加的。例如，以下 page 指令将同时导入 java.util.ArrayList 和 java.util.Date 类型。

```
<%@page import="java.util.ArrayList"%>
<%@page import="java.util.Date"%>
```

如下写法的效果一样：

```
<%@page import="java.util.ArrayList, java.util.Date"%>
```

一个 page 指令可以同时有多个属性。下面的代码设定了 session 属性和 buffer 属性。

```
<%@page session="false" buffer="16kb"%>
```

D.4.2　include 指令

可以使用 include 指令将其他文件中的内容包含到当前 JSP 页面。一个页面中可以有多个 include 指令。若存在一个内容会在多个不同页面中使用或一个页面不同位置使用的情况，则将该内容模块化到一个 include 文件会非常有用。

include 指令的语法如下：

```
<%@ include file="url"%>
```

其中，@和 include 间的空格不是必需的，URL 为被包含文件的相对路径，若 URL 以一个斜杠（/）开始，则该 URL 为文件在服务器上的绝对路径，否则为当前 JSP 页面的相对路径。

JSP 转换器处理 include 指令时，将指令替换为指令所包含文件的内容。换句话说，若存在清单 D.4 的 copyright.jspf 文件，以及主文件清单 D.5 的 main.jsp 页面：

清单 D.4　copyright.jspf 包含文件

```
<hr/>
&copy;2015 Brainy Software Inc.
<hr/>
```

清单 D.5　main.jsp 页面

```
<!DOCTYPE html>
<html>
<head><title>Including a file</title></head>
<body>
This is the included content: <hr/>
<%@ include file="copyright.jspf"%>
</body>
</html>
```

则在 main.jsp 页面中应用 include 指令和编写如下页面的效果是一样的：

```
<!DOCTYPE html>
<html>
<head><title>Including a file</title></head>
<body>
This is the included content: <hr/>
<hr/>
&copy;2015 Brainy Software Inc.
<hr/>
</body>
</html>
```

如上示例中，为保证 include 指令能正常工作，copyright.jspf 文件必须同 main.jsp 位于相同的目录。按照惯例，以 JSPF 为扩展名的文件代表 JSP fragement。虽然 JSP fragement 现在被称为 JSP segment，但为保证一致性，JSPF 后缀名依然被保留。

注意，include 指令也可以包含静态 HTML 文件。

此外，include 动作（类似于 include 指令）会在 D.6 一节讨论。理解两者之间的区别非常重要，具体细微的差别参见 D.6 节的解释。

D.5　脚本元素

一个脚本程序是一个 java 代码块，以<%符号开始，以%>符号结束。以清单 D.6 的 scriptletTest.jsp 页面为例：

清单 D.6　使用脚本程序（scriptletTest.jsp）

```
<%@page import="java.util.Enumeration"%>
<!DOCTYPE html>
<html>
<head><title>Scriptlet example</title></head>
<body>
<b>Http headers:</b><br/>
<%-- first scriptlet --%>
<%
    for (Enumeration<String> e = request.getHeaderNames();
            e.hasMoreElements(); ){
        String header = e.nextElement();
        out.println(header + ": " + request.getHeader(header) +
                "<br/>");
    }
```

```
    String message = "Thank you.";
%>
<hr/>
<%-- second scriptlet --%>
<%
    out.println(message);
%>
</body>
</html>
```

在上述清单 D.6 的 JSP 页面中，有两个脚本程序，需要注意的是定义在一个脚本程序中的变量可以被其后续的脚本程序使用。

脚本程序的第一行代码可以紧接<%标记，最后一行代码也可以紧接%>标记，不过，这会降低代码的可读性。

D.5.1 表达式

每个表达式都会被 JSP 容器执行，并使用隐式对象 out 的打印方法输出结果。表达式以“<%=”开始，并由“%>”结束。例如，在下面一行文本中，黑体字为一个表达式。

```
Today is <%=java.util.Calendar.getInstance().getTime()%>
```

注意，表达式无需分号结尾。

JSP 容器首先执行 java.util.Calendar.getInstance().getTime()，并将计算结果传递给 out.print()，这与如下脚本程序的效果一样：

```
Today is
<%
    out.print(java.util.Calendar.getInstance().getTime());
%>
```

D.5.2 声明

可以声明能在 JSP 页面中使用的变量和方法。声明以“<%!”开始，并以“%>”结束。例如，清单 D.7 的 declarationTest.jsp 页面展示了一个 JSP 页面，该页面声明了一个名为 getTodaysDate 的方法。

清单 D.7 使用声明（declarationTest.jsp）

```
<%!
    public String getTodaysDate() {
        return new java.util.Date();
```

```
    }
%>
<!DOCTYPE html>
<html>
<head><title>Declarations</title></head>
<body>
Today is <%=getTodaysDate()%>
</body>
</html>
```

在 JSP 页面中，一个声明可以出现在任何地方，并且一个页面可以有多个声明。

可以使用声明来重写 JSP 页面，实现类的 init 和 destroy 方法。通过声明 jspInit 方法，来重写 init 方法。通过声明 jspDestroy 方法，来重写 destory 方法。这两种方法说明如下。

- jspInit。这种方法类似于 javax.servlet.Servlet 的 init 方法。JSP 页面在初始化时调用 jspInit。不同于 init 方法，jspInit 没有参数。还可以通过隐式对象 config 访问 ServletConfig 对象。
- jspDestroy。这种方法类似于 Servlet 的 destroy 方法，在 JSP 页面将被销毁时调用。

清单 D.8 呈现的 lifeCycle.jsp 页面演示了如何重写 jspInit 和 jspDestroy。

清单 D.8 lifeCycle.jsp 页面

```
<%!
    public void jspInit() {
        System.out.println("jspInit ...");
    }
    public void jspDestroy() {
        System.out.println("jspDestroy ...");
    }
%>
<!DOCTYPE html>
<html>
<head><title>jspInit and jspDestroy</title></head>
<body>
Overriding jspInit and jspDestroy
</body>
</html>
```

lifeCycle.jsp 页面会被转换成如下的 Servlet:

```
package org.apache.jsp;
import javax.servlet.*;
import javax.servlet.http.*;
```

```
import javax.servlet.jsp.*;

public final class lifeCycle_jsp extends
        org.apache.jasper.runtime.HttpJspBase
        implements org.apache.jasper.runtime.JspSourceDependent {

    public void jspInit() {
        System.out.println("jspInit ...");
    }

    public void jspDestroy() {
        System.out.println("jspDestroy ...");
    }

    private static final javax.servlet.jsp.JspFactory _jspxFactory =
            javax.servlet.jsp.JspFactory.getDefaultFactory();

    private static java.util.Map<java.lang.String,java.lang.Long>
       _jspx_dependants;

    private javax.el.ExpressionFactory _el_expressionfactory;
    private org.apache.tomcat.InstanceManager _jsp_instancemanager;

    public java.util.Map<java.lang.String,java.lang.Long>
            getDependants() {
        return _jspx_dependants;
    }

    public void _jspInit() {
        _el_expressionfactory =
                _jspxFactory.getJspApplicationContext(
                getServletConfig().getServletContext())
                .getExpressionFactory();
        _jsp_instancemanager =
                org.apache.jasper.runtime.InstanceManagerFactory
                .getInstanceManager(getServletConfig());
    }

    public void _jspDestroy() {
    }

    public void _jspService(final
            javax.servlet.http.HttpServletRequest request, final
            javax.servlet.http.HttpServletResponse response)
            throws java.io.IOException,
            javax.servlet.ServletException {
```

```
        final javax.servlet.jsp.PageContext pageContext;
        javax.servlet.http.HttpSession session = null;
        final javax.servlet.ServletContext application;
        final javax.servlet.ServletConfig config;
        javax.servlet.jsp.JspWriter out = null;
        final java.lang.Object page = this;
        javax.servlet.jsp.JspWriter _jspx_out = null;
        javax.servlet.jsp.PageContext _jspx_page_context = null;

        try {
            response.setContentType("text/html");
            pageContext = _jspxFactory.getPageContext(this, request,
                response, null, true, 8192, true);
            _jspx_page_context = pageContext;
            application = pageContext.getServletContext();
            config = pageContext.getServletConfig();
            session = pageContext.getSession();
            out = pageContext.getOut();
            _jspx_out = out;

            out.write("\n");
            out.write("<!DOCTYPE html>\n");
            out.write("<html>\n");
            out.write("<head><title>jspInit and jspDestroy" +
                    "</title></head>\n");
            out.write("<body>\n");
            out.write("Overriding jspInit and jspDestroy\n");
            out.write("</body>\n");
            out.write("</html>");
        } catch (java.lang.Throwable t) {
            if (!(t instanceof
                    javax.servlet.jsp.SkipPageException)){
                out = _jspx_out;
                if (out != null && out.getBufferSize() != 0)
                    try {
                        out.clearBuffer();
                    } catch (java.io.IOException e) {
                    }
                if (_jspx_page_context != null)
                    _jspx_page_context.handlePageException(t);
            }
        } finally {
            _jspxFactory.releasePageContext(_jspx_page_context);
        }
    }
}
```

注意生成的 Servlet 类中的 jspInit 和 jspDestroy 方法。

现在可以通过如下 URL 访问 lifeCycle.jsp:

```
http://localhost:8080/jspdemo/lifeCycle.jsp
```

第一次访问页面时，可以在控制台上看到“jspInit...”，并且在 Servlet/JSP 容器关闭时看到“jspDestory...”。

D.5.3　禁用脚本元素

随着 JSP 2.0 对表达式语言的加强，推荐的做法是：在 JSP 页面中用 EL 访问服务器端对象且不写 Java 代码。因此，从 JSP 2.0 起，可以通过在部署描述符中的<jsp-property-group>定义一个 scripting-invalid 元素，来禁用脚本元素。

```
<jsp-config>
    <jsp-property-group>
        <url-pattern>*.jsp</url-pattern>
        <scripting-invalid>true</scripting-invalid>
    </jsp-property-group>
</jsp-config>
```

D.6　动作

动作是第三种类型的语法元素，它们被转换成 Java 代码来执行操作，如访问一个 Java 对象或调用方法。本节仅讨论所有 JSP 容器支持的标准动作。除了标准之外，还可以创建自定义标签来执行某些操作。

下面是一些标准的动作。

D.6.1　useBean

useBean 将创建一个关联 Java 对象的脚本变量。这是早期分离表示层和业务逻辑的努力之一。随着其他技术的发展，如自定义标签和表达语言，现在很少使用 useBean 方式。

清单 D.9 的 useBeanTest.jsp 页面是一个示例，它创建一个 java.util.Date 实例，并赋值给名为 today 的脚本变量，然后在表达式中使用它。

清单 D.9　useBeanTest.jsp 页面

```
<!DOCTYPE html>
<html>
```

```
<head>
    <title>useBean</title>
</head>
<body>
<jsp:useBean id="today" class="java.util.Date"/>
<%=today%>
</body>
</html>
```

在 Tomcat 中，上述代码会被转换为如下代码：

```
java.util.Date today = null;
today = (java.util.Date) _jspx_page_context.getAttribute("today",
        javax.servlet.jsp.PageContext.REQUEST_SCOPE);
if (today == null) {
    today = new java.util.Date();
    _jspx_page_context.setAttribute("today", today,
            javax.servlet.jsp.PageContext.REQUEST_SCOPE);
}
```

访问这个页面，会输出当前的日期和时间。

D.6.2 setProperty 和 getProperty

setProperty 动作可对一个 Java 对象设置属性，而 getProperty 则会输出 Java 对象的一个属性。清单 D.11 中的 getSetPropertyTest.jsp 页面展示了如何设置和输出在清单 D.10 中定义的 Employee 类实例的 firstName 属性。

清单 D.10 Employee 类

```
package jspdemo;
public class Employee {
    private String id;
    private String firstName;
    private String lastName;

    public String getId() {
        return id;
    }
    public void setId(String id) {
        this.id = id;
    }
    public String getFirstName() {
        return firstName;
    }
```

```
    public void setFirstName(String firstName) {
        this.firstName = firstName;
    }
    public String getLastName() {
        return lastName;
    }
    public void setLastName(String lastName) {
        this.lastName = lastName;
    }
}
```

清单 D.11 getSetPropertyTest.jsp 页面

```
<!DOCTYPE html>
<html>
<head>
<title>getProperty and setProperty</title>
</head>
<body>
<jsp:useBean id="employee" class="jspdemo.Employee"/>
<jsp:setProperty name="employee" property="firstName" value="Abigail"/>
First Name: <jsp:getProperty name="employee" property="firstName"/>
</body>
</html>
```

D.6.3 include

include 动作用来动态地引入另一个资源。可以引入另一个 JSP 页面，也可以引入一个 Servlet 或一个静态的 HTML 页面。例如，清单 D.12 的 jspIncludeTest.jsp 页面使用 include 动作来引入 menu.jsp 页面。

清单 D.12 jspIncludeTest.jsp 页面

```
<!DOCTYPE html>
<html>
<head>
<title>Include action</title>
</head>
<body>
<jsp:include page="jspf/menu.jsp">
    <jsp:param name="text" value="How are you?"/>
</jsp:include>
</body>
</html>
```

这里，理解 include 指令和 include 动作非常重要。对于 include 指令，资源引入发生在页面转换时，即当 JSP 容器把页面转换为生成的 Servlet 时。而 include 动作，资源引入发生在请求页面时。因此，使用 include 动作是可以传递参数的，而 include 指令不支持。

第二个不同是，include 指令对引入的文件扩展名不做特殊要求。但 include 动作，若引入的文件需以 JSP 页面处理，则其文件扩展名必须是 JSP。若使用 jspf 为扩展名，则该页面被当作静态文件。

D.6.4 forward

forward 动作把当前页面转向到其他资源。下面的代码将从当前页转向到 login.jsp 页面。

```
<jsp:forward page="jspf/login.jsp">
    <jsp:param name="text" value="Please login"/>
</jsp:forward>
```

D.7 错误处理

JSP 提供了很好的错误处理能力。除了在 Java 代码中可以使用 try 语句，还可以指定一个特殊页面——当应用页面遇到未捕获的异常时，用户将看到一个精心设计的网页来解释发生了什么，而不是用户无法理解的一条错误信息。

请使用 page 指令的 isErrorPage 属性（属性值必须为 True）来标识一个 JSP 页面是错误页面。清单 D.13 展示了一个错误处理程序。

清单 D.13 errorHandler.jsp 页面

```
<%@page isErrorPage="true"%>
<!DOCTYPE html><html>
<head><title>Error</title></head>
<body>
An error has occurred. <br/>
Error message:
<%
    out.println(exception.toString());
%>
</body>
</html>
```

其他需要防止未捕获的异常的页面必须使用 page 指令的 errorPage 属性，来指向错误处理页面。例如，清单 D.14 中的 buggy.jsp 页面就使用了清单 D.13 的错误处理程序。

清单 D.14　buggy.jsp 页面

```
<%@page errorPage="errorHandler.jsp"%>
Deliberately throw an exception
<%
    Integer.parseInt("Throw me");
%>
```

运行的 buggy.jsp 页面会抛出一个异常。不过，我们不会看到由 Servlet/JSP 容器生成的错误消息。相反，会看到 errorHandler.jsp 页面的内容。

D.8　小结

JSP 是构建在 Java Web 应用程序的第二种技术，是 Servlet 技术的补充，而不是 Servlet 技术的替代。一个精心设计的 Java Web 应用程序会同时使用 Servlet 和 JSP。

在本附录中，我们学习了 JSP 是如何工作的，以及如何编写 JSP 页面。现在，我们已经知道 JSP 的隐式对象，并能在 JSP 页面使用 3 个语法元素：指令、脚本元素和动作。

附录 E 部署描述符

部署一个 Servlet 3 或 Servlet 3.1 应用程序是一件轻而易举的事。通过 Servlet 注解类型，对于不太复杂的应用程序，可以部署没有描述符的 Servlet/JSP 应用程序。尽管如此，在需要更加精细配置的情况下，仍然需要部署描述符。首先，部署描述符必须被命名为 web.xml 并且位于 WEB-INF 目录下，Java 类必须放置在 WEB-INF/classes 目录下，而 Java 类库则必须位于 WEB-INF/lib 目录下。所有的应用程序资源必须包装成一个以 war 为后缀的 jar 文件。

本附录会讨论部署和部署描述符，这是一个应用程序的重要组成部分。

E.1 概述

在 Servlet 3 之前，部署工作必然涉及部署描述符，即 web.xml 文件，我们在该文件中配置应用程序的各个方面。但在 Servlet 3 中，部署描述符是可选的，因为我们可以使用标注来映射一个 URL 模式的资源。不过，若存在如下场景，则依然需要部署描述符。

- 需要传递初始参数给 ServletContext。
- 有多个过滤器，并要指定调用顺序。
- 需要更改会话超时设置。
- 要限制资源的访问，并配置用户身份验证方式。

清单 E.1 展示了部署描述符的框架。它必须被命名为 web.xml 且合并在应用目录的 WEB-INF 目录下。

清单 E.1 部署描述符的框架

```
<?xml version="1.0" encoding="ISO-8859-1"?>
<web-app version="3.1" xmlns="http://xmlns.jcp.org/xml/ns/javaee"
```

```
    xmlns:xsi="http://www.w3.org/2001/XMLSchema-instance"
    xsi:schemaLocation="http://xmlns.jcp.org/xml/ns/javaee
➥ http://xmlns.jcp.org/xml/ns/javaee/web-app_3_1.xsd"
     [metadata-complete=" true|false" ]
>

    ...

</web-app>
```

xsi:schemaLocation 属性指定了模式文档的位置，以便可以进行验证。version 属性指定了 Servlet 规范的版本。

可选的 metadata-complete 属性指定部署描述符是否是完整的，若值为 True，则 Servlet/JSP 容器将忽略 Servlet 注解。若值为 False 或不存在，则容器必须检查类文件的 Servlet 注解，并扫描 web fragments 文件。

web-app 元素是文档的根元素，并且可以具有如下子元素：

- Servlet 声明。
- Servlet 映射。
- ServletContext 的初始化参数。
- 会话配置。
- 监听器类。
- 过滤器定义和映射。
- MIME 类型映射。
- 欢迎文件列表。
- 错误页面。
- JSP 特定的设置。
- JNDI 设置。

每个元素的配置规则可见 web-app_3_1.xsd 文档，可以从如下网站下载：

```
http://xmlns.jcp.org/xml/ns/javaee/web-app_3_1.xsd
```

web-app_3_1.xsd 包括另一种模式（webcommon_3_1.xsd），其中包含了大部分信息。可从如下网站下载：

```
http://xmlns.jcp.org/xml/ns/javaee/web-common_3_1.xsd
```

webcommon_3_1.xsd 包括以下两种模式：

- javaee_7.xsd，定义了其他 Java 共享公共元素 EE7 的部署类型（EAR、JAR 和 RAR）。
- jsp_2_7.xsd，定义要素配置的一部分，JSP 根据 JSP 2.3 规范中的应用。

本节列出了在部署描述符中常见的 Servlet 和 jsp 元素，但不包括那些不在 Servlet 或 JSP 规范中的 Java EE 元素。

E.1.1 核心元素

本节将详细介绍各重要元素的细节。web-app 的子元素可以以任何顺序出现。某些元素，如 session-config、jsp-config 和 login-config 只能出现一次，而另一些元素，如 Servlet、filter 和 welcome-file-list 可以出现很多次。

后续几个小节会分别描述在<web-app>元素下的一级元素。若要查找非<web-app>下的非一级元素，请查找其父元素。例如，taglib 元素在“jsp-config”下，而 load-on-startup 在 Servlet 下。本节后续小节按字母顺序排序。

E.1.2 context-param

可用 context-param 元素传值给 ServletContext。这些值可以被任何 Servlet/JSP 页面读取。context-param 元素由名称/值对构成，并可以通过调用 ServletContext 的 getInitParameter 方法来读取。可以定义多个 context-param 元素，每个参数名在本应用中必须唯一。ServletContext.getInitParameterNames()方法会返回所有的参数名称。

每个 context-param 元素必须包含一个 param-name 元素和一个 param-value 元素。param-name 定义参数名，而 param-value 定义参数值。另有一个可选的元素，即 description 元素，用来描述参数。

下面是 context-param 元素的两个例子。

```
<context-param>
    <param-name>location</param-name>
    <param-value>localhost</param-value>
</context-param>
<context-param>
    <param-name>port</param-name>
    <param-value>8080</param-value>
```

```
    <description>The port number used</description>
</context-param>
```

E.1.3　distributable

若定义了 distributable 元素，则表明应用程序已部署到分布式的 Servlet/JSP 容器。distributable 元素必须是空的。例如，下面是一个 distributable 例子。

```
<distributable/>
```

E.1.4　error-page

error-page 元素包含一个HTTP错误代码与资源路径或Java异常类型与资源路径之间的映射关系。error-page 元素定义容器在特定 HTTP 错误或异常时应返回的资源路径。

error-page 元素由如下成分构成：

- error-code，指定一个 HTTP 错误代码。
- exception-type，指定 Java 的异常类型（全路径名称）。
- location，指定要被显示的资源位置。该元素必须以“/”开始。

下面的配置告诉 Servlet/JSP 容器，当出现 HTTP 404 时，显示位于应用目录下的 error.html 页面。

```
<error-page>
    <error-code>404</error-code>
    <location>/error.html</location>
</error-page>
```

下面的配置告诉 Servlet/JSP 容器，当发生 ServletException 时，显示 exception.html 页面。

```
<error-page>
    <exception-type>javax.servlet.ServletException</exception-type>
    <location>/exception.html</location>
</error-page>
```

E.1.5　filter

filter 指定一个 Servlet 过滤器。该元素至少包括一个 filter-name 元素和一个 filter-class 元素。此外，它也可以包含以下元素：icon、display-name、discription、init-param 以及 async-supported。

filter-name 元素定义了过滤器的名称。过滤器名称必须全局唯一。filter-class 元素指定过滤器类的全路径名称。可由 init-param 元素来配置过滤器的初始参数（类似于<context-param>），一个过滤器可以有多个 init-param。

下面是 Upper Case Filter 和 Image Filter 这两个 filter 元素。

```
<filter>
    <filter-name>Upper Case Filter</filter-name>
    <filter-class>com.example.UpperCaseFilter</filter-class>
</filter>
<filter>
    <filter-name>Image Filter</filter-name>
    <filter-class>com.example.ImageFilter</filter-class>
    <init-param>
        <param-name>frequency</param-name>
        <param-value>1909</param-value>
     </init-param>
    <init-param>
        <param-name>resolution</param-name>
        <param-value>1024</param-value>
    </init-param>
</filter>
```

E.1.6 filter-mapping

过滤器映射元素是指定过滤器要被映射到的一个或多个资源。过滤器可以被映射到 Servlet 或者 URL 模式。将过滤器映射到 Servlet 会致使过滤器对该 Servlet 产生作用。将过滤器映射到 URL 模式，则会使其对所有 URL 与该 URL 模式匹配的资源进行过滤。过滤的顺序与过滤器映射元素在部署描述符中的顺序一致。

过滤器映射元索中包含一个 filter-name 元素和一个 url-pattern 元素或者 servlet-name 元素。

filter-name 元素的值必须与利用 filter 元素声明的某一个过滤器名称相匹配。

下面的例子中是两个过滤器元素和两个过滤器映射元素。

```
<filter>
    <filter-name>Logging Filter</filter-name>
    <filter-class>com.example.LoggingFilter</filter-class>
</filter>
<filter>
    <filter-name>Security Filter</filter-name>
    <filter-class>com.example.SecurityFilter</filter-class>
</filter>
```

```
<filter-mapping>
    <filter-name>Logging Filter</filter-name>
    <servlet-name>FirstServlet</servlet-name>
</filter-mapping>
<filter-mapping>
    <filter-name>Security Filter</filter-name>
    <url-pattern>/*</url-pattern>
</filter-mapping>
```

E.1.7　listener

listener 元素用来注册一个侦听器。其子元素 listener-class 包含监听器类的全路径名。如下是一个示例：

```
<listener>
    <listener-class>com.example.AppListener</listener-class>
</listener>
```

E.1.8　locale-encoding-mapping-list 和 locale-encoding-mapping

locale-encoding-mapping-list 元素包含了一个或多个 locale-encoding-mapping 元素。每个 locale-encoding-mapping 定义了 locale 以及编码的映射，分别用 locale 以及 encoding 元素定义。locale 元素的值必须是在 ISO 639 中定义的语言编码，如 en，或者是采用"语言编码_国家编码"格式，如 en_US。其中，国家编码值必须在 ISO 3166 中定义。

如下是一个示例：

```
<locale-encoding-mapping-list>
    <locale-encoding-mapping>
        <locale>ja</locale>
        <encoding>Shift_JIS</encoding>
    </locale-encoding-mapping>
</locale-encoding-mapping-list>
```

E.1.9　login-config

login-config 元素包括 auth-method、realm-name 以及 form-login-config 元素，每个元素都是可选的。

auth-method 元素定义了认证方式，可选值为 BASIC、DIGEST、FORM 和 CLIENT-CERT。

realm-name 元素定义了用于 BASIC 以及 DIGEST 认证方式的 realm 名称。

form-login-config 则定义了用于 FORM 认证方式的登录页面和失败页面。若没有采用 FORM 认证方式，则该元素被忽略。

form-login-config 元素包括 form-login-page 和 form-error-page 两个子元素。其中，form-login-page 配置了显示登录页面的资源路径，路径为应用目录的相对路径，且必须以“/”开始。form-error-page 则配置了登录失败时显示错误页面的资源路径。同样，路径为应用目录的相对路径，且必须以“/”开始。

下面是一个示例：

```
<login-config>
    <auth-method>DIGEST</auth-method>
    <realm-name>Members Only</realm-name>
</login-config>
```

另一个示例如下：

```
<login-config>
    <auth-method>FORM</auth-method>
    <form-login-config>
        <form-login-page>/loginForm.jsp</form-login-page>
        <form-error-page>/errorPage.jsp</form-error-page>
    </form-login-config>
</login-config>
```

E.1.10　mime-mapping

mime-mapping 元素用来把一个 MIME 类型映射到一个扩展名。该元素由一个 extension 元素和一个 mime-type 元素组成。示例如下：

```
<mime-mapping>
    <extension>txt</extension>
    <mime-type>text/plain</mime-type>
</mime-mapping>
```

E.1.11　security-constraint

security-constraint 元素允许对一组资源进行限制访问。

security-constraint 元素有如下子元素：一个可选的 display-name 元素、一个或多个 web-resource-collection 元素、可选的 auth-constraint 元素和一个可选的 user-data-constraint 元素。

web-resource-collection 元素标识了一组需要进行限制访问的资源集合。这里，你可以定义

URL模式和所限制的HTTP方法。如果没有定义HTTP方法，则表示应用于所有HTTP方法。

auth-constraint元素指明哪些角色可以访问受限制的资源集合。如果没有指定，则应用于所有角色。

user-data-constraint元素用于指示在客户端和Servlet/JSP容器传输的数据是否保护。

web-resource-collection元素包含一个web-resource-name元素、一个可选的description元素、零个或多个url-pattern元素，以及零个或多个http-method元素。

web-resource-name元素指定受保护的资源名称。

http-method元素指定HTTP方法，如GET、POST或TRACE。

auth-constraint元素包含一个可选的description元素、零个或多个role-name元素。role-name元素指定角色名称。

user-data-constraint元素包含一个可选的description元素和一个transport-guarantee元素。transport-guarantee元素的取值范围如下：NONE、INTEGRAL或CONFIDENTIAL。NONE表示该应用程序不需要安全传输保障。INTEGRAL意味着服务器和客户端之间的数据在传输过程中不能被篡改。CONFIDENTIAL意味着必须加密传输数据。大多数情况下，安全套接字层（SSL）会应用于INTEGRAL或CONFIDENTIAL。

下面是一个例子：

```
<security-constraint>
    <web-resource-collection>
        <web-resource-name>Members Only</web-resource-name>
        <url-pattern>/members/*</url-pattern>
    </web-resource-collection>
    <auth-constraint>
        <role-name>payingMember</role-name>
    </auth-constraint>
</security-constraint>

<login-config>
    <auth-method>Digest</auth-method>
    <realm-name>Digest Access Authentication</realm-name>
</login-config>
```

E.1.12 security-role

security-role元素声明用于安全限制的安全角色。这个元素有一个可选的description元素和 role-name 元素。下面是一个例子：

```
<security-role>
    <role-name>payingMember</role-name>
</security-role>
```

E.1.13 Servlet

Servlet 元素用来配置 Servlet，包括如下子元素：

- 一个可选的 icon 元素。
- 一个可选的 description 元素。
- 可选的 display-name 元素。
- 一个 servlet-name 元素。
- 一个 servlet-class 元素或一个 jsp-file 元素。
- 零个或更多的 init-param 元素。
- 一个可选的 load-on-startup 元素。
- 可选的 run-as 元素。
- 可选的 enabled 元素。
- 可选的 async-supported 元素。
- 可选的 multipart-config 元素。
- 零个或多个 security-role-ref 元素。

一个 Servlet 元素至少必须包含一个 servlet-name 元素和一个 servlet-class 元素，或者一个 servlet-name 元素和一个 jsp-file 元素。

servlet-name 元素定义的 Servlet 名称在应用程序中必须是唯一的。

servlet-class 元素指定的类名为全路径名。

jsp-file 元素指定 JSP 页面的路径，该路径是应用程序的相对路径，必须以“/”开始。

init-param 的子元素可以用来传递一个初始参数给 Servlet。init-param 元素的构成同 context-param。

可以使用 load-on-startup 元素在当 Servlet/JSP 容器启动时自动加载 Servlet。加载一个

Servlet 是指实例化 Servlet 和调用它的 init 方法。使用此元素可以避免由于加载 Servlet 而导致对第一个请求的响应延迟。如果该元素指定了 jsp-file 元素，则 JSP 文件被预编译成 Servlet，并加载该 Servlet。

load-on-startup 可以指定用一个整数值来指定加载顺序。例如，如果有两个 Servlet 且都包含一个 load-on-startup 元素，则值小的 Servlet 优先加载。若没有指定值或值为负数，则由 Web 容器决定如何加载。若两个 Servlet 具有相同的 load-on-startup 值，则加载 Servlet 的顺序不能确定。

run-as 用来覆盖调用 EJB 的安全标识。角色名是当前 Web 应用程序定义的安全角色之一。

security-role-ref 元素将在调用 Servlet 的 isUserInRole 方法时角色名映射到应用程序定义的安全角色。security-role-ref 元素包含一个可选的 description 元素、一个 role-name 元素和一个 role-link 元素。

role-link 元素用于将安全角色映射到一个已定义的安全角色，它必须包含一个在 security-role 元素中定义的安全角色。

async-supported 元素是一个可选的元素，其值可以是 True 或 False。它表示 Servlet 是否支持异步处理。

enabled 元素也是一个可选的元素，它的值可以是 True 或 False。设置此元素为 False，则禁用这个 Servlet。

例如，映射安全角色“PM”与角色名字“payingMember”的配置如下：

```
<security-role-ref>
    <role-name>PM</role-name>
    <role-link>payingMember</role-link>
</security-role-ref>
```

这样，若属于 payingMember 角色的用户调用 Servlet 的 isUserInRole（“payingMember”）方法，则结果为真。

下面是 Servlet 元素的两个例子：

```
<servlet>
    <servlet-name>UploadServlet</servlet-name>
    <servlet-class>com.brainysoftware.UploadServlet</servlet-class>
    <load-on-startup>10</load-on-startup>
</servlet>
<servlet>
```

```
    <servlet-name>SecureServlet</servlet-name>
    <servlet-class>com.brainysoftware.SecureServlet</servlet-class>
    <load-on-startup>20</load-on-startup>
</servlet>
```

E.1.14 servlet-mapping

servlet-mapping 元素将一个 Servlet 映射到一个 URL 模式。该元素必须有一个 servlet-name 元素和 url-pattern 元素。

下面的 servlet-mapping 元素映射一个 Servlet 到/first。

```
<servlet>
    <servlet-name>FirstServlet</servlet-name>
    <servlet-class>com.brainysoftware.FirstServlet</servlet-class>
</servlet>
<servlet-mapping>
    <servlet-name>FirstServlet</servlet-name>
    <url-pattern>/first</url-pattern>
</servlet-mapping>
```

E.1.15 session-config

session-config 元素定义了用于 javax.servlet.http.HttpSession 实例的参数。此元素可包含一个或更多的以下内容：session-timeout、cookie-config 或 tracking-mode。

session-timeout 元素指定会话超时间隔（分钟）。该值必须是整数。如果该值是零或负数，则会话将永不超时。

cookie-config 元素定义了跟踪会话创建的 cookie 的配置。

tracking-mode 元素定义了跟踪会话模式，其有效值是 COOKIE、URL 或 SSL。

下面定义的 session-config 元素使得应用的 HttpSession 对象在 12 分钟不活动后失效。

```
<session-config>
    <session-timeout>12</session-timeout>
</session-config>
```

E.1.16 welcome-file-list

welcome-file-list 元素指定当用户在浏览器中输入的 URL 不包含一个 Servlet 名称或 JSP 页面或静态资源时所显示的文件或 Servlet。

welcome-file-list 元素包含一个或多个 welcome-file 元素。welcome-file 元素包含默认的文件名。如果在第一个 welcome-file 元素指定的文件没有找到，则 Web 容器将尝试显示第二个，直到最后一个。

下面是一个 welcome-file-list 元素的例子。

```
<welcome-file-list>
    <welcome-file>index.htm</welcome-file>
    <welcome-file>index.html</welcome-file>
    <welcome-file>index.jsp</welcome-file>
</welcome-file-list>
```

下面的示例，第一个 welcome-file 元素指定了一个在应用程序目录下的 index.html；第二个 welcome-file 为 servlet 目录下的欢迎文件。

```
<welcome-file-list>
    <welcome-file>index.html</welcome-file>
    <welcome-file>servlet/welcome</welcome-file>
</welcome-file-list>
```

E.1.17　JSP-Specific Elements

<web-app>元素下的 jsp-config 元素，可以指定 JSP 配置。它可以具有零个或多个标签 taglib 和零个或多个 jsp-property-group 元素。下面首先介绍 taglib 元素，然后介绍 jsp-property-group 元素。

E.1.18　taglib

taglib 元素定义了 JSP 定制标签库。taglib 元素包含一个 taglib-uri 元素和 taglib-location 元素。taglib-uri 元素定义了 Servlet/JSP 应用程序所用的标签库的 URI，其值相当于在部署描述符路径。

taglib-location 元素指定 TLD 文件的位置。

下面是一个 taglib 元素的例子。

```
<jsp-config>
    <taglib>
        <taglib-uri>
            http://brainysoftware.com/taglib/complex
        </taglib-uri>
        <taglib-location>/WEB-INF/jsp/complex.tld
    </taglib-location>
```

```
  </taglib>
</jsp-config>
```

E.1.19 jsp-property-group

jsp-property-group 中的元素可为一组 JSP 文件统一配置属性。使用<jsp-property-group>子元素可做到以下几点：

- 指示 EL 显示是否忽略。
- 指示脚本元素是否允许。
- 指明页面的编码信息。
- 指示一个资源是 JSP 文件（XML 编写）。
- 预包括和代码自动包含。

jsp-property-group 包含如下子元素：

- 一个可选的 description 元素。
- 一个可选的 display-name 元素。
- 一个可选的 icon 元素。
- 一个或多个 url-pattern 元件。
- 一个可选的 el-ignored 元素。
- 一个可选的 page-encoding 元素。
- 一个可选的 scripting-invalid 元素。
- 一个可选的 is-xml 元素。
- 零个或多个 include-prelude 元素。
- 零个或多个 include-code 元素。

url-pattern 元素用来指定可应用相应属性配置的 URL 模式。

el-ignored 元素值为 True 或 False。True 值表示在匹配 URL 模式的 JSP 页面中，EL 表达式无法被计算，该元素的默认值是 False。

page-encoding 元素指定 JSP 页面的编码。page-encoding 的有效值和页面的 pageEncoding

的有效值相同。若 page-encoding 指定值与匹配 URL 模式的 JSP 页面中的 pageEncoding 属性值不同时，则会产生一个转换时错误。同样，若 page-encoding 指定值与 XML 文档声明的编码不同，也会产生一个转换时错误。

scripting-invalid 元素值为 True 或 False。True 值是指匹配 URL 模式的 JSP 页面不支持<% scripting %>语法。scripting-invalid 元素的默认值是 False。

is-xml 元素值为 True 或 False，True 表示匹配 URL 模式的页面是 JSP 文件。

include-prelude 元素值为相对于 Servlet/JSP 应用的相对路径。若设置该元素，则匹配 URL 模式的 JSP 页面开头处会自动包含给定路径文件（同 include 指令）。

include-coda 元素值为相对于 Servlet/JSP 应用的相对路径。若设置该元素，则匹配 URL 模式的 JSP 页面结尾处会自动包含给定路径文件（同 include 指令）。

在下面的例子中，jsp-property-group 配置所有的 JSP 页面无法执行 EL 表达式。

```
<jsp-config>
    <jsp-property-group>
        <url-pattern>*.jsp</url-pattern>
        <el-ignored>true</el-ignored>
    </jsp-property-group>
</jsp-config>
```

在下面的例子中，jsp-property-group 配置所有的 JSP 页面不支持<% scripting %>语法。

```
<jsp-config>
    <jsp-property-group>
        <url-pattern>*.jsp</url-pattern>
        <scripting-invalid>true</scripting-invalid>
    </jsp-property-group>
</jsp-config>
```

E.2 部署

从 Servlet 1.0 开始，可以很方便地部署一个 Servlet/JSP 应用程序。仅需要将应用原始目录结构压缩成一个 WAR 文件。可以在 JDK 中使用 jar 工具或使用流行的工具，如 WinZip。需要确保压缩文件有.war 扩展名。如果使用 WinZip，则在压缩完成后重命名文件。

war 文件必须包含所有库文件、类文件、HTML 文件、JSP 页面、图像文件以及版权声明（如果有的话）等，但不包括 Java 源文件。任何人都可以获取一个 war 文件的副本，并部署到一个 Servlet/JSP 容器上。

E.3 Web Fragment

Servlet 3 添加了 web fragment 特性，用来为已有的 Web 应用部署插件或框架。web fragment 被设计成部署描述符的补充，而无需编辑 web.xml 文件。一个 Web fragment 基本上包含了常用的 Web 对象，如 Servlet，过滤器和监听器，其他资源，如 JSP 页面和静态图像的包文件（jar 文件）。一个 web fragment 也可以有一个描述符，类似的部署描述符的 XML 文档。web fragment 描述符必须命名为 web-fragment.xml，并位于包的 META-INF 目录下。一个 web fragment 描述符可能包含可出现在部署描述符 web-app 元素下的任意元素，再加上一些 web fragment 的特定元素。一个应用程序可以有多个 web fragment。

清单 E.2 显示了 web fragment 描述符，以黑体形式突出显示与部署描述符之间的不同内容。在 web fragment 的根元素必须是 web-fragment 元素，其可以有 metadata-complete 属性。如果 metadata-complete 属性的值为 True，则包含在 web fragment 中所有的类注释将被跳过。

清单 E.2 web fragment.xml 文件的框架

```
<?xml version="1.0" encoding="ISO-8859-1"?>
<web-fragment version="3.1" xmlns="http://xmlns.jcp.org/xml/ns/javaee"
    xmlns:xsi="http://www.w3.org/2001/XMLSchema-instance"
    xsi:schemaLocation="http://xmlns.jcp.org/xml/ns/javaee
➥ http://xmlns.jcp.org/xml/ns/javaee/web-fragment_3_1.xsd"
     [metadata-complete="true|false"]
>

    ...

</web-fragment>
```

作为一个例子，在 fragmentdemo 应用程序中包含的 fragment.jar 文件是一个 web fragment。该 jar 文件已经导入到 WEB-INF/lib 目录下。本实例的重点不在于 fragmentdemo，而是 web fragment 项目。该项目包含一个 Servlet（fragment.servlet.FragmentServlet，见清单 E.3）和 webfragment.xml 文件（见清单 E.4）。

清单 E.3 FragmentServlet 类

```
package fragment.servlet;
import java.io.IOException;
```

```
import java.io.PrintWriter;
import javax.servlet.ServletException;
import javax.servlet.http.HttpServlet;
import javax.servlet.http.HttpServletRequest;
import javax.servlet.http.HttpServletResponse;

public class FragmentServlet extends HttpServlet {

    private static final long serialVersionUID = 940L;

    public void doGet(HttpServletRequest request, HttpServletResponse
       response)
            throws ServletException, IOException {

        response.setContentType("text/html");
        PrintWriter out = response.getWriter();
        out.println("A plug-in");
    }
}
```

清单 E.4　webfragment.xml 文件

```
<?xml version="1.0" encoding="ISO-8859-1"?>
<web-fragment version="3.1" xmlns="http://xmlns.jcp.org/xml/ns/javaee"
    xmlns:xsi="http://www.w3.org/2001/XMLSchema-instance"
    xsi:schemaLocation="http://xmlns.jcp.org/xml/ns/javaee
➥ http://xmlns.jcp.org/xml/ns/javaee/web-fragment_3_1.xsd">

    <servlet>
        <servlet-name>FragmentServlet</servlet-name>
        <servlet-class>fragment.servlet.FragmentServlet</servletclass>
    </servlet>
    <servlet-mapping>
        <servlet-name>FragmentServlet</servlet-name>
        <url-pattern>/fragment</url-pattern>
    </servlet-mapping>
</web-fragment>
```

FragmentServlet 是发送一个字符串到浏览器的一个简单的 Servlet。web-fragment.xml 文件注册并映射该 Servlet。fragment.jar 文件结构如图 E.1 所示。

```
fragment
  servlet
    FragmentServlet.class
META-INF
  web-fragment.xml
```

图 E.1　fragment.jar 文件结构

使用如下 URL 调用测试该 Servlet。

```
http://localhost:8080/fragmentdemo/fragment
```

可以看到 Fragment Servlet 的输出。

E.4 小结

本附录介绍了如何配置和部署 Servlet/JSP 应用程序。本附录首先介绍一个典型应用的目录结构，然后详细阐释了部署描述符。

发布一个应用程序有两种方式：一种是以目录结构的形式；另一种是打包成一个单一的 war 文件进行部署。

特 别 优 惠

购买本书的读者专享异步社区购书优惠券。

使用方法：注册成为社区用户，在下单购书时输入 S4XC5 使用优惠码，然后点击“使用优惠码”，即可在原折扣基础上享受全单9折优惠。（订单满39元即可使用，本优惠券只可使用一次）

纸电图书组合购买

社区独家提供纸质图书和电子书组合购买方式，价格优惠，一次购买，多种阅读选择。

社区里还可以做什么？

提交勘误

您可以在图书页面下方提交勘误，每条勘误被确认后可以获得100积分。热心勘误的读者还有机会参与书稿的审校和翻译工作。

写作

社区提供基于 Markdown 的写作环境，喜欢写作的您可以在此一试身手，在社区里分享您的技术心得和读书体会，更可以体验自出版的乐趣，轻松实现出版的梦想。

如果成为社区认证作译者，还可以享受异步社区提供的作者专享特色服务。

会议活动早知道

您可以掌握 IT 圈的技术会议资讯，更有机会免费获赠大会门票。

加入异步

扫描任意二维码都能找到我们：

异步社区	微信服务号	微信订阅号	官方微博	QQ 群：436746675

社区网址：www.epubit.com.cn

投稿 & 咨询：contact@epubit.com.cn